Control of Multiphase Machines and Drives

Control of Multiphase Machines and Drives

Special Issue Editors

Federico Barrero
Ignacio González-Prieto

MDPI • Basel • Beijing • Wuhan • Barcelona • Belgrade

Special Issue Editors

Federico Barrero
University of Seville
Spain

Ignacio González-Prieto
University of Malaga
Spain

Editorial Office
MDPI
St. Alban-Anlage 66
4052 Basel, Switzerland

This is a reprint of articles from the Special Issue published online in the open access journal *Energies* (ISSN 1996-1073) in 2020 (available at: https://www.mdpi.com/journal/energies/special_issues/multiphase_machines_drives).

For citation purposes, cite each article independently as indicated on the article page online and as indicated below:

LastName, A.A.; LastName, B.B.; LastName, C.C. Article Title. *Journal Name* **Year**, *Article Number*, Page Range.

ISBN 978-3-03928-136-7 (Pbk)
ISBN 978-3-03928-137-4 (PDF)

Cover image courtesy of Ignacio Gonzalez Prieto.

© 2020 by the authors. Articles in this book are Open Access and distributed under the Creative Commons Attribution (CC BY) license, which allows users to download, copy and build upon published articles, as long as the author and publisher are properly credited, which ensures maximum dissemination and a wider impact of our publications.

The book as a whole is distributed by MDPI under the terms and conditions of the Creative Commons license CC BY-NC-ND.

Contents

About the Special Issue Editors vii

Angel Gonzalez-Prieto, Ignacio Gonzalez-Prieto, Mario J. Duran and Federico Barrero
Efficient Model Predictive Control with Natural Fault-Tolerance in Asymmetrical Six-Phase Induction Machines
Reprinted from: *Energies* **2019**, *12*, 3989, doi:10.3390/en12203989 1

Cristina Martin, Federico Barrero, Manuel R. Arahal and Mario J. Duran
Model-Based Predictive Current Controllers in Multiphase Drives Dealing with Natural Reduction of Harmonic Distortion
Reprinted from: *Energies* **2019**, *12*, 1679, doi:10.3390/en12091679 17

Yassine Kali, Magno Ayala and Jorge Rodas
Predictive-Fixed Switching Current Control Strategy Applied to Six-Phase Induction Machine
Reprinted from: *Energies* **2019**, *12*, 2294, doi:10.3390/en12122294 29

Agnieszka Kowal, Manuel R. Arahal, Cristina Martin and Federico Barrero
Constraint Satisfaction in Current Control of a Five-Phase Drive with Locally Tuned Predictive Controllers
Reprinted from: *Energies* **2019**, *12*, 2715, doi:10.3390/en12142715 43

Daniel R. Ramirez, Cristina Martin, Agnieszka Kowal G. and Manuel R. Arahal
Min-Max Predictive Control of a Five-Phase Induction Machine
Reprinted from: *Energies* **2019**, *12*, 3713, doi:10.3390/en12193713 52

Federico Barrero, Mario Bermudez, Mario J. Duran, Pedro Salas and Ignacio Gonzalez-Prieto
Assessment of a Universal Reconfiguration-less Control Approach in Open-Phase Fault Operation for Multiphase Drives
Reprinted from: *Energies* **2019**, *12*, 4698, doi:10.3390/en12244698 61

Antoine Cizeron, Javier Ojeda, Eric Labouré and Olivier Béthoux
Prediction of PWM-Induced Current Ripple in Subdivided Stator Windings Using Admittance Analysis
Reprinted from: *Energies* **2019**, *12*, 4418, doi:10.3390/en12234418 73

Jose Riveros, Joel Prieto, Marco Rivera, Sergio Toledo and Raúl Gregor
A Generalised Multifrequency PWM Strategy for Dual Three-Phase Voltage Source Converters
Reprinted from: *Energies* **2019**, *12*, 1398, doi:10.3390/en12071398 92

Yassine Kali, Magno Ayala and Jorge Rodas
Current Control of a Six-Phase Induction Machine Drive Based on Discrete-Time Sliding Mode with Time Delay Estimation
Reprinted from: *Energies* **2019**, *12*, 170, doi:10.3390/en12010170 105

Daniel Gutierrez-Reina, Federico Barrero, Jose Riveros, Ignacio Gonzalez-Prieto, Sergio L. Toral and Mario J. Duran
Interest and Applicability of Meta-Heuristic Algorithms in the Electrical Parameter Identification of Multiphase Machines
Reprinted from: *Energies* **2019**, *12*, 314, doi:10.3390/en12020314 122

About the Special Issue Editors

Federico Barrero (M 04; SM 05) received his MSc and PhD degrees in Electrical and Electronic Engineering from the University of Seville, Spain, in 1992 and 1998, respectively. In 1992, he joined the Electronic Engineering Department at the University of Seville, where he is currently Full Professor. He received the Best Paper Award from the IEEE Trans. on Ind. Electron. for 2009 and IET Electric Power Applications for 2010-2011.

Ignacio González Prieto (Doctor of Engineering). Ignacio González Prieto was born in Malaga, Spain, in 1987. He received his Industrial Engineer and MSc degrees in fluid mechanics from the University of Malaga, Malaga, Spain, in 2012 and 2013, respectively, and PhD degree in Electronic Engineering from the University of Seville, Seville, Spain, in 2016. His research interests include multiphase machines, wind energy systems, and electrical vehicles.

Article

Efficient Model Predictive Control with Natural Fault-Tolerance in Asymmetrical Six-Phase Induction Machines

Angel Gonzalez-Prieto [1], Ignacio Gonzalez-Prieto [1], Mario J. Duran [1] and Federico Barrero [2,*]

1. Department of Electrical Engineering, University of Malaga, 29071 Malaga, Spain; anggonpri@gmail.com (A.G.-P.); ignaciogp87@gmail.com (I.G.-P.); mjduran@uma.es (M.J.D)
2. Department of Electrical Engineering, University of Seville, 41092 Seville, Spain
* Correspondence: fbarrero@us.es; Tel.: +34-666-38-97-37

Received: 9 September 2019; Accepted: 18 October 2019; Published: 20 October 2019

Abstract: Multiphase machines allow enhancing the performance of wind energy conversion systems from the point of view of reliability and efficiency. The enhanced robustness has been traditionally achieved with a mandatory post-fault control reconfiguration. Nevertheless, when the regulation of x-y currents in multiphase drives is done in open-loop mode, the reconfiguration can be avoided. As a consequence, the reliability of the system increases because fault detection errors or delays have no impact on the post-fault performance. This capability has been recently defined as natural fault tolerance. From the point of view of the efficiency, multiphase machines present a better power density than three-phase machines and lower per-phase currents for the same voltage rating. Moreover, the implementation of control strategies based on a variable flux level can further reduce the system losses. Targeting higher reliability and efficiency for multiphase wind energy conversion systems, this work proposes the implementation of an efficient model predictive control using virtual voltage vectors for six-phase induction machines. The use of virtual voltage vectors allows regulation of the x-y currents in open-loop mode and achieving the desired natural fault tolerance. Then, a higher efficiency can be achieved with a simple and universal cost function, which is valid both in pre- and post-fault situations. Experimental results confirm the viability and goodness of the proposal.

Keywords: model predictive control; multiphase induction machines; natural fault tolerance

1. Introduction

Promoted by new energy policies in different countries, renewable energies currently play an important role in the electricity market [1]. There is a wide variety of clean energies aiming to replace fossil fuels, but wind energy is the most installed one in the world. In fact, a high percentage of the demanded electric energy is obtained nowadays from wind energy conversion systems (WECS). This high penetration of wind energy has, in turn, increased the requirements from transmission system operators to WECS.

In this context, the efficiency and reliability are two desirable features for the newly developed WECSs. Since multiphase machines present enhanced fault-tolerant (FT) capabilities and better power densities than conventional three-phase systems [2–4], multiphase generators appear as promising candidates in full-power WECS. The improved post-fault operation capability of multiphase machines can provide economic benefits when the corrective maintenance tasks are complex, as it is the case in offshore locations [5]. Furthermore, the better power density of multiphase machines allows lower copper losses, increasing the wind resource exploitation. Hence, multiphase machines have become an interesting alternative in modern wind farms [6].

In terms of reliability, the FT capability, and particularly the open-phase fault (OPF) operation, can be considered the most appreciated advantage of multiphase machines, but this higher reliability has been traditionally achieved with a mandatory OPF localization/isolation [7] and a post-fault control reconfiguration [8]. Nevertheless, the natural FT recently suggested in [9] circumvents this standard trend of post-fault control strategies and provides simpler means to obtain an improved reliability. The model predictive control (MPC) based on virtual voltage vectors (VV) from [9] provides an enhanced post-fault capability without fault localization and control reconfiguration. The implementation of VVs in MPC allows regulating the x-y currents in open-loop mode. This fact avoids the conflict of the α-β and x-y controllers when a new restriction appears in the system due to the OPF. As a consequence, the VV-MPC becomes a universal control strategy for pre- and post-fault situations, this being a highly attractive feature for the wind energy industry. However, regardless of the selected post-fault strategy, the integrity of the system in post-fault situation can only be preserved by decreasing the torque/power production (derating) [8,10].

Efficiency is also a highly appreciated characteristic of WECS because lower losses imply a better exploitation of the wind energy resources and greater torque/power production in post-fault situations. The concept of efficient control is based on the reduction of the magnetic flux in the machine at light loads to minimize the copper losses. Efficient control has been traditionally implemented using two different strategies: search control (SC) [11–13] and loss model control (LMC) [14–16] methods. SC algorithms produce an online perturbation in the magnetic flux when looking for the optimal balance between torque and magnetization. Then, the convergence provided by SC methods is slow, although it is insensitive to machine parameters [13]. On the other hand, LMC techniques are based on the theoretical estimation of the magnetic flux using a model of the system [14], arising higher speed convergences and sensitivity to variations in the machine parameters. In order to develop more competitive WECSs, an efficient LMC strategy was successfully implemented in [16] for a six-phase IM in post-fault situation using field-oriented control (FOC). Nevertheless, the developed efficient model was based on a selected topology (with parallel converters) and specific post-fault situations, i.e., this efficient model is only valid for these particular conditions. If the machine is driven with a single VSC supply and the system is healthy, the efficient control must be revisited to achieve a more general strategy that ensures its validity both before and after the OPF occurrence.

With the aim of providing natural fault tolerance capability and lower losses, an efficient MPC technique based on virtual voltage vectors (EVV-MPC) was suggested in [17]. This work confirmed for the first time the interest of EVV-MPC, improving the efficiency and reliability of six-phase IM drives in pre- and post-fault situations while maintaining the same control scheme. However, the analysis presented in [17] was only supported by simulations. This work goes beyond, providing a detailed analysis and experimental results that confirm the EVV-MPC features in steady or transient states and healthy or post-fault situations and avoiding any sensitivity to parameter detuning when evaluating the optimal magnetic flux level. The paper has been structured as follows: A description of the analyzed multiphase induction machine is included in the next section, and the model predictive control using virtual voltage vectors and the concept of natural fault tolerance are detailed in Section 3. Next, the efficient control scheme presented in this work is detailed in Section 4 and validated through experimentation in Section 5. The obtained conclusions are finally summarized in the last section.

2. Asymmetrical Six-Phase Induction Drives

The studied multiphase drive is formed by an asymmetrical six-phase induction machine with distributed windings where two independent and isolated neutral points are created, supplied by two three-phase and two-level voltage source converters (VSCs) connected to a single DC link (Figure 1). The switching state of every VSC leg can be defined using a binary variable S_{ij}, being $S_{ij} = 0$ if the lower switch is ON and the upper switch is OFF, and $S_{ij} = 1$ if the opposite situation occurs. According to the number of phases of the proposed IM drive, $2^6 = 64$ switching states exist. It is common to group theses switching states in a vector $[S] = \{S_{a1}, S_{b1}, S_{c1}, S_{a2}, S_{b2}, S_{c2}\}$ that determines the obtained stator phase voltage from the DC-link voltage (V_{dc}) as follows:

$$[M] = \frac{V_{DC}}{3} \cdot \begin{bmatrix} 2 & -1 & -1 & 0 & 0 & 0 \\ -1 & 2 & -1 & 0 & 0 & 0 \\ -1 & -1 & 2 & 0 & 0 & 0 \\ 0 & 0 & 0 & 2 & -1 & -1 \\ 0 & 0 & 0 & -1 & 2 & -1 \\ 0 & 0 & 0 & -1 & -1 & 2 \end{bmatrix} \cdot [S]^T, \quad (1)$$

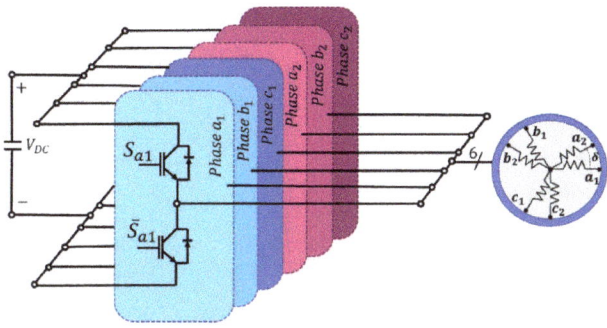

Figure 1. Asymmetrical six-phase IM ($\delta = 30°$) drive topology.

The current invariant Clarke transformation is usually applied to simplify the mathematical description of the system from phase-variables into two orthogonal stationary subspaces, α-β and x-y (see Figure 2):

$$[T] = \frac{1}{3} \begin{bmatrix} 1 & -1/2 & -1/2 & \sqrt{3}/2 & -\sqrt{3}/2 & 0 \\ 0 & \sqrt{3}/2 & -\sqrt{3}/2 & 1/2 & 1/2 & -1 \\ 1 & -1/2 & -1/2 & -\sqrt{3}/2 & \sqrt{3}/2 & 0 \\ 0 & -\sqrt{3}/2 & \sqrt{3}/2 & 1/2 & 1/2 & -1 \\ 1 & 1 & 1 & 0 & 0 & 0 \\ 0 & 0 & 0 & 1 & 1 & 1 \end{bmatrix}, c \quad (2)$$

$$\begin{bmatrix} v_{\alpha s} v_{\beta s} v_{xs} v_{ys} v_{0+} v_{0-} \end{bmatrix}^T = [T] \cdot [v_{a1} v_{b1} v_{c1} v_{a2} v_{b2} v_{c2}]^T,$$

where α-β components are related with the flux/torque generation, x-y components produce copper losses, and the isolated neutral points simplify the analysis and avoid triples stator current harmonics. Based on this fact, the vector space decomposition (VSD) approach is usually applied to define the six-phase IM as follows [18]:

$$\begin{aligned} v_{\alpha s} &= \left(R_s + L_s \cdot \tfrac{d}{dt}\right) i_{\alpha s} + M \cdot \tfrac{di_{\alpha r}}{dt}, \\ v_{\beta s} &= \left(R_s + L_s \cdot \tfrac{d}{dt}\right) \cdot i_{\beta s} + M \cdot \tfrac{di_{\beta r}}{dt}, \\ v_{xs} &= \left(R_s + L_{ls} \cdot \tfrac{d}{dt}\right) \cdot i_{xs}, \\ v_{ys} &= \left(R_s + L_{ls} \cdot \tfrac{d}{dt}\right) \cdot i_{ys}, \\ 0 &= \left(R_r + L_r \cdot \tfrac{d}{dt}\right) i_{\alpha r} + M \cdot \tfrac{di_{\alpha s}}{dt} + \omega_r \cdot L_r \cdot i_{\beta r} + \omega_r \cdot M \cdot i_{\beta s}, \\ 0 &= \left(R_r + L_r \cdot \tfrac{d}{dt}\right) \cdot i_{\beta r} + M \cdot \tfrac{di_{\beta s}}{dt} - \omega_r \cdot L_r \cdot i_{\alpha r} - \omega_r \cdot M \cdot i_{\alpha s}, \\ T_e &= p \cdot M \cdot (i_{\beta r} \cdot i_{\alpha s} - i_{\alpha s} \cdot i_{\beta s}), \end{aligned} \quad (3)$$

where $L_s = L_{ls} + 3 \cdot L_m$, $L_r = L_{lr} + 3 \cdot L_m$, $M = 3 \cdot L_m$, $\omega_r = p \cdot \omega_m$, with p and ω_m being the mechanical speed. In addition, indices s and r denote stator and rotor variables and subscripts l and m denote leakage and magnetizing inductance, respectively.

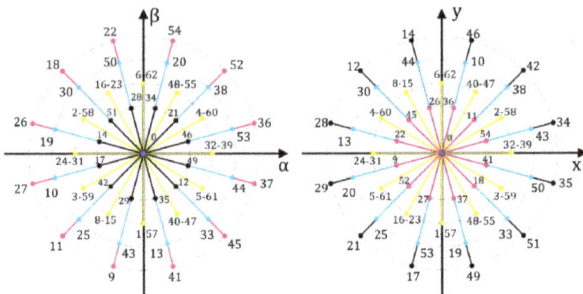

Figure 2. Voltage vectors in α-β (left plot) and x-y (right plot) subspaces for an asymmetrical six-phase IM drive.

To simplify the control a synchronous reference frame (d-q) can be employed where the d-component is related with flux production and the q- component with the torque production. This transformation of the reference frame is obtained applying the Park transformation in Equation (4) to the α-β components as follows:

$$[D] = \begin{bmatrix} \cos(\theta_s) & \sin(\theta_s) \\ -\sin(\theta_s) & \cos(\theta_s) \end{bmatrix},$$
$$[i_{ds} i_{qs}]^T = [D] \cdot [i_{\alpha s} i_{\beta s}]^T, \quad [i_{x1s} i_{y1s}]^T = [D] \cdot [i_{xs} i_{ys}]^T, \tag{4}$$

with θ_s being the angle of the reference frame, obtained from the measured speed (ω_m) and the estimated slip [19].

3. VV-MPC with Natural Fault Tolerance

Standard MPC (see Figure 3) uses an outer speed control loop with a PI controller to obtain the reference value of the q-current, whereas the d-current is usually set to a fixed value that is proportional to the rated flux in the base-speed region. In addition, an inner predictive current controller regulates the power converter that feeds the machine and commands the electromechanical system. This current controller is based on a discretized machine model that is used to predict currents in future operation points [9]. The predicted currents are then included in a predefined cost function (Equation (5)) where different error terms are included and whose final value is minimized to find an optimal switching state:

$$J_1 = K_1 \cdot e_{qs}^2 + K_2 \cdot e_{ds}^2 + K_3 \cdot e_{xs}^2 + K_4 \cdot e_{ys}^2, \tag{5}$$

where K_i coefficients, also called weighting factors and defined to achieve regulation goals and drive features, multiply i-error components that are defined in this case as follows:

$$\begin{aligned} e_{qs} &= (i_{qs}^*|_{k+2} - \hat{i}_{qs}|_{k+2}), & e_{xs} &= (i_{xs}^*|_{k+2} - \hat{i}_{xs}|_{k+2}), \\ e_{ds} &= (i_{ds}^*|_{k+2} - \hat{i}_{ds}|_{k+2}), & e_{ys} &= (i_{ys}^*|_{k+2} - \hat{i}_{ys}|_{k+2}), \end{aligned} \tag{6}$$

being the predicted currents in $k+2$ ("$\hat{i}_{qs}|_{k+2}$") compared with the reference currents ("$i_{qs}^*|_{k+2}$") in $k+2$ to apply the optimal switching state to the power converter.

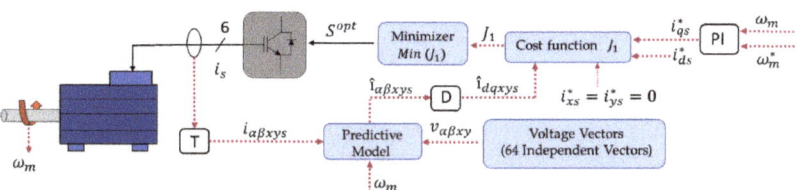

Figure 3. Standard MPC scheme for six-phase induction motor drives.

Note that the reference values of the *x-y* currents are usually set to a null value in order to reduce stator copper losses in distributed-winding machines. However, it was recently stated that its performance can be highly degraded if the stator leakage inductance of the machine presents a low value [19]. The standard MPC technique applies a single switching state during the whole sampling time, generating simultaneously voltage vectors in α-β and *x-y* planes when active voltage vectors are applied. Consequently, *x-y* currents flow through the machine, spoiling the current quality and increasing the stator copper losses.

To solve the aforementioned disadvantage, [20,21] proposed the implementation of virtual voltage vectors for MPC strategies. The VVs are created taking advantage of the special localization of the available six-phase IM voltage vectors. As shown in Figure 2, voltage vectors can be classified depending on their magnitude in small, medium, medium-large and large voltage vectors where medium-large and large vectors that share their direction in the α-β plane have opposite directions in the *x-y* plane. Hence, it is possible to obtain a virtual voltage vector as a combination of medium-large and large vectors, providing a null average *x-y* voltage production. For this purpose, it is necessary to apply different times for medium-large and large voltage vectors. In a six-phase VSC, the application time of each vector must be $t_1 = 0.73 \cdot T_m$ (for large voltage vectors) and $t_2 = 0.27 \cdot T_m$ (for medium-large voltage vectors), being T_m the sampling period. Following this approach, 12 active virtual voltage vectors can be defined as:

$$VV_i = t_1 \cdot V_{large} + t_2 \cdot V_{medium-large} \tag{7}$$

With the application of these VV, the control of *x-y* currents is performed in open-loop mode, with no inclusion of these components into the control strategy. This simplification results in a reduced predictive model that skip the *x-y* equations [20,21]. Consequently, the number of weighting factors can be reduced compared to standard MPC and the *x-y* term can be eliminated from the cost function as follows:

$$J_2 = K_1 \cdot \left(i_{qs}^*|_{k+2} - \hat{i}_{qs}|_{k+2} \right)^2 + K_2 \cdot \left(i_{ds}^*|_{k+2} - \hat{i}_{ds}|_{k+2} \right)^2, \tag{8}$$

Figure 4 shows the VV-MPC scheme where the main differences with the standard MPC (Figure 3) have been colored in magenta. Additionally, VVs provide MPC with a natural fault-tolerant characteristic because the regulation of the *x-y* currents is performed in open-loop mode and the controllers' conflicts after the fault occurrence are eliminated [9]. Based on this, the post-fault reconfiguration can be suppressed, and the reliability of the entire system is improved thanks to a reduced impact of the fault detection errors or delays on the post-fault operation and performance.

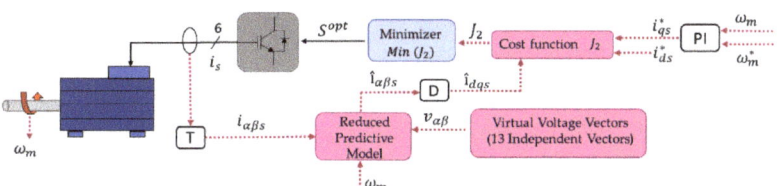

Figure 4. VV-MPC scheme for six-phase induction motor drives.

This natural fault tolerance concept has introduced a new paradigm in the field of fault-tolerant multiphase drives, where a mandatory post-fault reconfiguration of the healthy control scheme has been historically used in the event of an OPF [22–25]. Conventional post-fault strategies reconfigure the control scheme according to the fault localization in order to avoid the controllers' conflicts. Since the fault situation is missed and the control action is not modified, the drive performance can be distorted because α-β and x-y subspaces are no longer independent and the controllers usually have conflicting objectives. If a single OPF in phase a_1 occurs in the analyzed system, the new restriction is [8]:

$$i_{xs} = -i_{\alpha s} \tag{9}$$

This restriction does not allow the independent regulation of stator currents in the primary and secondary planes, forcing a conflict if healthy controllers are applied in post-fault operation. In order to avoid any conflict and preserve the controllability of the system in post-fault operation, x-y reference currents have been traditionally modified in order to drive α-β and x-y controllers into a single direction (with $i_{xs}^* = -i_{\alpha s}^*$ [8]). However, if the x-y currents are regulated in the open-loop mode, the conflict is automatically avoided because the x-y terms are not included in the cost function [9]. Nevertheless, and regardless of the selected post-fault strategy, the machine must be derated in order to safeguard the integrity of the system. This fact could promote the selection of efficient strategies where the magnetic flux level is variated at light loads to reduce the stator copper losses.

4. Efficient Model Predictive Control Based on Virtual Voltage Vectors

From the efficiency point of view, the most widely selected approach is to adapt the flux to reduce the stator copper losses. Focusing on these losses, they can be expressed as the product of stator resistance and the squared RMS (Root Mean Square) value of phase currents:

$$P_{cu} = 6 \cdot R_s \cdot i_s^2, \tag{10}$$

On the other hand, according to the VSD approach and considering the Park transformation, the RMS value of the phase currents can be calculated as:

$$i_s = \sqrt{i_{ds}^2 + i_{qs}^2 + i_{xs}^2 + i_{ys}^2}, \tag{11}$$

Trefore, to obtain the minimum stator copper losses, Equation (11) must be minimized. Efficient strategies are based on the magnetic flux variation according to the operating point with the aim of minimizing stator copper losses. The operation point of an induction machine can be usually defined by the magnetic flux level and the required electromagnetic torque, being expressed the electromagnetic torque in IMs with distributed windings using d-q currents as follows:

$$T_e = \frac{P \cdot L_m^2}{L_r} \cdot i_{ds} \cdot i_{qs}, \tag{12}$$

Based on Equations (10)–(12), a constrained minimization problem can be formulated in order to obtain the minimum RMS value of the phase currents for each operation point as:

$$\begin{aligned} \text{Minimize}: & \ i_{ds}^2 + i_{qs}^2 + i_{xs}^2 + i_{ys}^2, \\ \text{subject to}: & \ T_e = \frac{P \cdot L_m^2}{L_r} \cdot i_{ds} \cdot i_{qs}, \end{aligned} \tag{13}$$

Although there are different solvers, the Lagrange function solves this constrained minimization problem in a straightforward manner as follows:

$$L(i_{ds}, i_{qs}, i_{xs}, i_{ys}, \lambda) = i_{ds}^2 + i_{qs}^2 + i_{xs}^2 + i_{ys}^2 + \lambda \cdot T_e - \frac{\lambda \cdot P \cdot L_m^2}{L_r} \cdot i_{ds} \cdot i_{qs}, \tag{14}$$

where λ is the Lagrange multiplier.

Implementing each minimization condition:

$$\begin{aligned}
\frac{\partial L}{\partial i_{ds}} &= 2 \cdot i_{ds} - \lambda \cdot \frac{P \cdot L_m^2}{L_r} \cdot i_{qs} = 0, \\
\frac{\partial L}{\partial i_{qs}} &= 2 \cdot i_{qs} - \lambda \cdot \frac{P \cdot L_m^2}{L_r} \cdot i_{ds} = 0, \\
\frac{\partial L}{\partial i_{xs}} &= 2 \cdot i_{xs} = 0, \\
\frac{\partial L}{\partial i_{ys}} &= 2 \cdot i_{ys} = 0, \\
\frac{\partial L}{\partial \lambda} &= T_e - \frac{P \cdot L_m^2}{L_r} \cdot i_{ds} \cdot i_{qs}
\end{aligned} \quad (15)$$

The reference value of the *d*-current that provides minimum stator current RMS value for a given load torque can be obtained from Equation (15) as:

$$i_{ds} = \sqrt{\frac{T_e \cdot L_r}{P \cdot L_m^2}}, \quad (16)$$

Note that Equation (15) also provides the value of *q* and *x-y* currents that achieve the minimum RMS value of phase currents for the corresponding electromagnetic torque. Even though *x-y* currents appear as a component in the calculation of stator phase current RMS values, they are not included in torque restriction equation, and do not affect the optimization problem, presenting a zero value as a solution. This fact has special relevance in faulty operation, where the *x-y* components have a non-null value due to the OPF constraint: since the maximum efficiency does not depend on the *x-y* currents, it is not necessary to modify the control scheme after the OPF occurrence and the control algorithm can maintain the efficiency in pre- and post-fault scenarios. On the other hand, i_{qs} can be obtained from the constrained minimization problem as:

$$i_{qs} = \sqrt{\frac{T_e \cdot L_r}{P \cdot L_m^2}}, \quad (17)$$

From a mathematical point of view, Equations (16) and (17) are identical, which means that it is possible to implement an efficient control scheme replacing the *d*-current rated value with the actual reference value of the *q*-current:

$$i_{ds}^* = i_{qs}^*, \quad (18)$$

However, this approach is not achievable if the electromagnetic torque imposed by the operating point is higher than a critical value (Equation (19)), since an overrated *d*-current leads to magnetic saturation and distorts the control performance.

$$T_e^{critical} = P \cdot \frac{L_m^2}{L_r} \cdot i_{ds}^{rated} \cdot i_{ds}^{rated} \quad (19)$$

The efficient strategy is implemented in VV-MPC with a new cost function (Equation (20)) that allows the electromagnetic torque production control (first term) and the magnetic flux adaptation to the corresponding operation point (second term). The method can be used when the system is operated with light load torques, which matches the specific derated condition in the post-fault situation to safeguard the integrity of the system.

$$J_3 = K_1 \cdot \left(i_{qs}^*|_{k+2} - \hat{i}_{qs}|_{k+2} \right)^2 + K_2 \cdot \left(i_{qs}^*|_{k+2} - \hat{i}_{ds}|_{k+2} \right)^2, \quad (20)$$

In summary, this new cost function does not only improve the efficiency of the system, but also maintains its natural fault-tolerant capability due to the open-loop regulation of *x-y* currents. Moreover, the introduced efficient cost function is also universal, being valid in pre- and post- fault situations and avoiding any control reconfiguration. Figure 5 shows the proposed EVV-MPC, where the introduced cost function is highlighted in yellow.

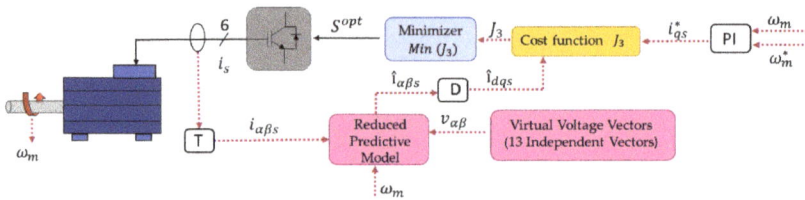

Figure 5. EVV-MPC scheme for six-phase induction motor drives.

5. Viability of the Proposal

The viability and goodness of the proposal is analyzed using the test bench shown in Figure 6, where the asymmetrical six-phase IM is fed by conventional two-level three-phase VSC (Semikron SKS22F modules (Semikron, Nuremberg, Germany)). The parameters of the custom-built six-phase IM have been obtained using ac-time domain and stand-still with inverter supply tests [26,27]. Table 1 shows the main parameters of the electric driver where the value of the stator resistance and leakage inductance of the α-β and x-y planes are the same.

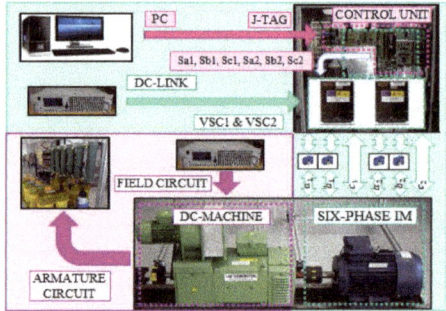

Figure 6. Experimental test bench.

Table 1. Electric drive parameters.

Power (kW)	0.8
Dc-link voltage(V)	300
Dead time (μs)	4
I_{peak}(A)	4.06
n_m(r/min)	1000
R_s(Ω)	4.2
R_r(Ω)	2
L_m(mH)	420
L_{ls}(mH)	1.5
L_{lr}(mH)	55

A single DC power supplies the VSC and the control actions are performed by a digital signal processor (TMS320F28335 from Texas Instruments, TI (Texas Instruments, Dallas, TX, USA)). The current and speed measurements are obtained using four hall-effect sensors (LEM LAH 25-NP (LEM, Bourg-la-Reine, France)) and a digital encoder (GHM510296R/2500 (Sensata, Attleboro, MA, USA)), respectively, while the six-phase IM is loaded by a coupled DC machine. Note that the armature of the DC machine is connected to a variable passive R load that dissipates the power and the load torque is consequently speed-dependent. Note also that the open-phase fault is forced from the TMS320F28335 using a controllable relay board between the inverter and the machine. Then, the performance of the proposed EVV-MPC can be assessed in pre- and post-fault situations, where four different tests have been designed.

A speed ramp test is firstly realized in pre-fault situation (see Figure 7) to evaluate the performance of the proposed controller (right column). The reference speed is changed from 200 to 400 rpm, and a VV-MPC strategy is also tested for comparison purposes using the same operation conditions (left column). The tracking of the reference speed is satisfactory regardless of the selected control strategy (see left and right columns, Figure 7a). However, the d-current in VV-MPC is constant all throughout the test (left column, Figure 7b), whereas in the case of EVV-MPC it follows the q-current reference according to the cost function detailed in Equation (20) (right column, Figure 7b). The difference between the reference of the q-current and the measured d-current is depicted in Figure 7c for both control strategies. As previously expressed in Section 4, a null difference reduces the RMS value of the phase currents, as it is shown in Figure 7e,f. Focusing on this issue, EVV-MPC presents an RMS phase current value of 0.57 A compared to 0.92 A in VV-MPC when the speed is low. This difference implies a reduction of 61.61% in the stator copper losses. When high-speed operation is reached, the RMS phase current value is 1.00 A using VV-MPC, whereas 0.60 A is obtained using EVV-MPC. A reduction of copper losses is clearly achieved using variable magnetic flux levels according to the operating point (see Figure 7g), since x-y currents have the same behavior using both control strategies in healthy situation (see Figure 7d). Based on these results in healthy operation, it can be confirmed the capability of the proposed EVV-MPC technique to reduce the copper losses for a similar switching frequency than VV-MPC as shown in Table 2.

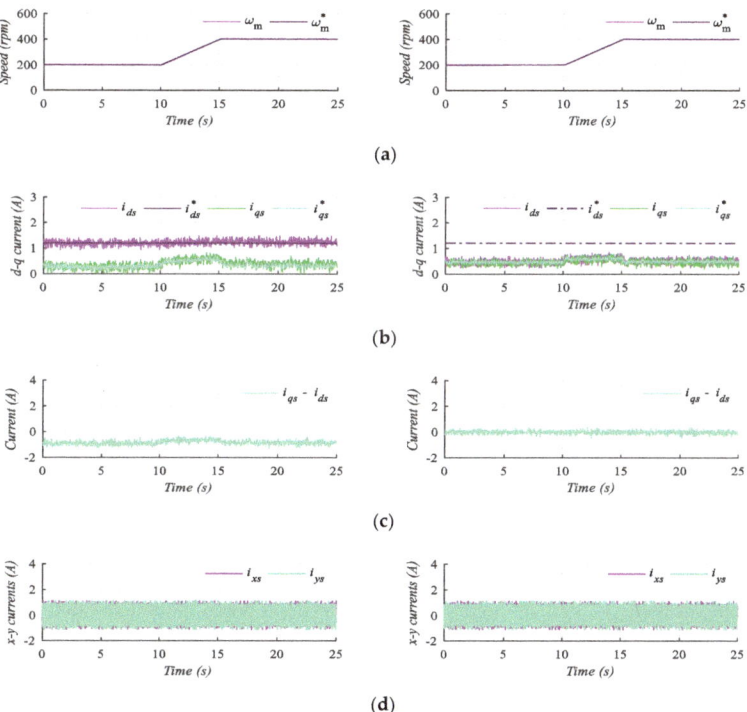

Figure 7. Cont.

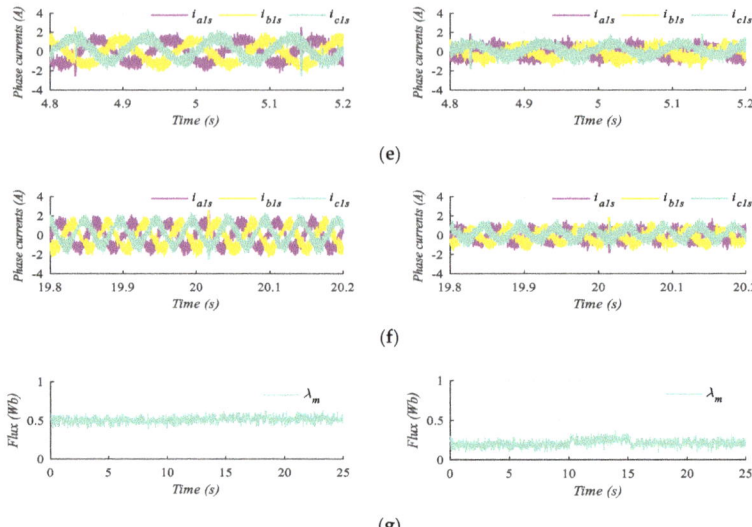

Figure 7. Pre-fault evaluation of the proposed VV-MPC technique (left column) versus EVV-MPC method (right column). From top to bottom: (**a**) Motor speed; (**b**) d-q currents; (**c**) the difference between q- and d-current; (**d**) x-y currents; (**e**) zoom 1 of set 1 of phase currents; (**f**) zoom 1 of set 2 of phase currents; and (**g**) stator magnetic flux.

Table 2. Switching frequency in pre-fault situation.

Speed	EVV-MPC	VV-MPC
200 rpm	4281 Hz	4296 Hz
400 rpm	4190 Hz	3925 Hz

A second test is conducted (see Figure 8) to justify the utilization of the proposed EVV-MPC technique (right column) in the post-fault situation. In this case, the performance in the transition from the pre- to post-fault situation is studied, and the VV-MPC technique is also evaluated (see left column) for the sake of comparison. The system is in healthy operation, but an open-phase fault in phase a_1 is forced at $t = 12.5$ s (Figure 8d). The current cannot flow through the open phase and a new restriction appears in the system ($i_x = -i_\alpha$) as it can be appreciated in Figure 8d,e. Note that the reference speed is satisfactorily tracked using the proposal without any control reconfiguration (see Figure 8a), since x-y currents are regulated in open-loop mode and any controller' conflict exists in post-fault situation. These results validate in fact the natural fault tolerance of MPC strategies when the regulation of x-y currents is realized in open-loop mode. On the other hand, the implemented cost function allows adapting the magnetic flux level in pre- and post-fault scenarios without any control reconfiguration, as expected. In this case, d-q currents maintain their reference values constant during the whole test (Figure 8b), since the operation point is constant. While the magnetic flux tracks the reference value in VV-MPC, a reduced magnetic level is applied in the case of EVV-MPC (Figure 8f) in order to lessen the stator copper losses. Focusing on the obtained phase currents, EVV-MPC strategy needs a lower value of the phase currents in pre- and post-fault situations to reach the same operation point than using VV-MPC. However, in the post-fault situation the reduction of the α-current also provokes a reduction of the x-current component since the α-β and x-y planes are no longer independent in faulty operation. Therefore, the implementation of the EVV-MPC strategy provides in the post-fault situation a reduction in the obtained copper losses due to the adaptation of the magnetic flux level. In order to quantify this efficiency improvement, Table 3 shows the relative reduction of RMS phase current values and the square of RMS phase current. This second term is related to the stator copper losses associated to theses currents in this test according to Equation (20), where the stator copper losses (SCL) can be obtained independently from the value of the stator resistance R_s as long as the stator resistances are equal in both tests. The obtained results confirm the goodness of the proposed controller in post-fault situation and validate the universality of the implemented cost-function in the pre- and post-fault situations.

$$\text{SCL (\%)} = \frac{P_{s\_VV} - P_{s\_EVV}}{P_{s\_VV}} \cdot 100 = \frac{6 \cdot R_s \cdot i^2_{phase\_VV} - 6 \cdot R_s \cdot i^2_{phase\_EVV}}{6 \cdot R_s \cdot i^2_{phase\_VV}} \cdot 100 = \frac{i^2_{phase\_VV} - i^2_{phase\_EVV}}{i^2_{phase\_VV}} \cdot 100 \quad (21)$$

Table 3. Relative reduction of the RMS phase current value and stator copper losses (SCL) of the healthy phase currents for VV-MPC and EVV-MPC in test 2.

Phase	%RMS$_{pre-fault}$	%RMS$_{post-fault}$	%SCL$_{pre-fault}$	%SCL$_{post-fault}$
b_1	41.23%	38.99%	65.44%	62.78%
c_1	40.93%	39.02%	65.10%	62.81%
a_2	40.40%	42.53%	64.48%	66.97%
b_2	40.93%	42.96%	65.10%	67.46%
c_2	39.40%	37.21%	63.28%	60.57%

The response of the proposed control scheme is finally evaluated in dynamic post-fault situations (see Figure 9). In this case, the VV-MPC method (left column) is again compared with the proposed EVV-MPC technique (right column), and an open-phase fault is forced in phase a_1 at the beginning of the test, appearing a new restriction in the system (Figure 9e,g). The speed is varied in a ramp-wise manner from 200 rpm to 400 rpm (see Figure 9a), being satisfactory the speed regulation in VV-MPC (left column) and EVV-MPC (right column). However, d-current changes to adapt its value to each operation point using the proposed EVV-MPC technique, adjusting also the magnetic flux level in dynamic post-fault situations. Meanwhile, the d-current value remains constant using VV-MPC during the whole test (Figure 9b), which only adjusts the q-current according to the operating point. Flux adaptation provides a reduction of α-β current amplitudes that are directly related, through inverse Clark transformation, with phase current amplitudes (Figure 9c,d). At low speed operating points, an RMS phase current value of 0.63 A is obtained using EVV-MPC versus 1.10 A with the VV-MPC technique, which implies a reduction of 67.20% of the stator copper losses. When high speed operating points are considered, an RMS phase value of 0.69 A is reached with EVV-MPC against 1.12 A using VV-MPC, which again implies an important reduction of stator copper losses (62.05% in the latter case). This test certifies the interest of the proposed control method that assures speed tracking and improves the efficiency in healthy and faulty operations, and in steady-state and transient conditions without introducing modifications in the control scheme.

A speed reversal test has been realized in pre-fault (left plots in Figure 10) and post-fault (right plots in Figure 10) situations when an open-phase fault occurs in phase a_1 at $t = 0$ s. It is shown in both cases that the speed and current tracking is satisfactory (Figure 10a,b), maintaining the efficient control both in healthy mode and in post-fault condition (Figure 10c). The x-y currents are regulated around zero in the pre-fault situation (Figure 10d, left column) and show the value $i_{xs} = -i_{\alpha s}$ that corresponds to the post-fault situation (Figure 10d, right column). As expected, phase currents change the sequence after the zero crossing (Figure 10e,f).

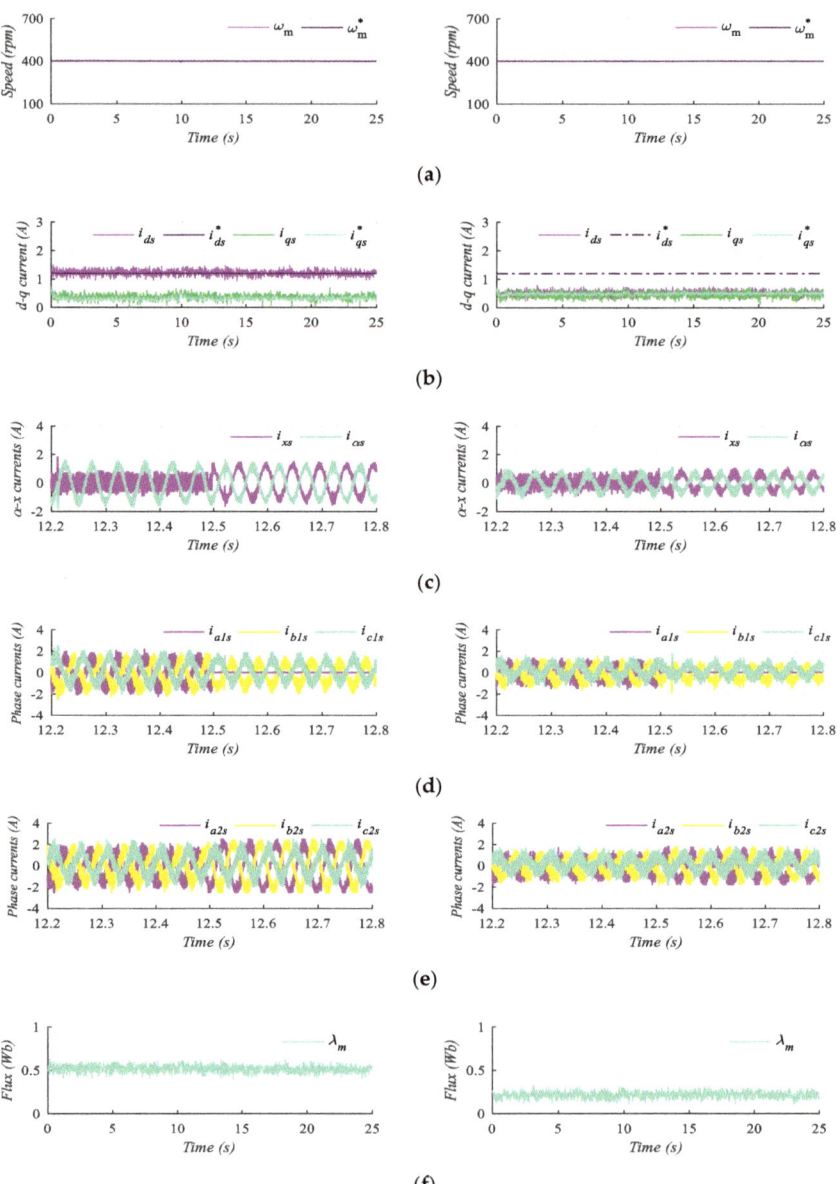

Figure 8. Trasition from pre to post-fault situations using VV-MPC (left column) and the proposed EVV-MPC (right column) control methods. From top to bottom: (**a**) Motor speed; (**b**) d-q currents; (**c**) x-α currents; (**d**) set 1 of phase currents (**e**) set 2 of phase currents; and (**f**) stator magnetic flux.

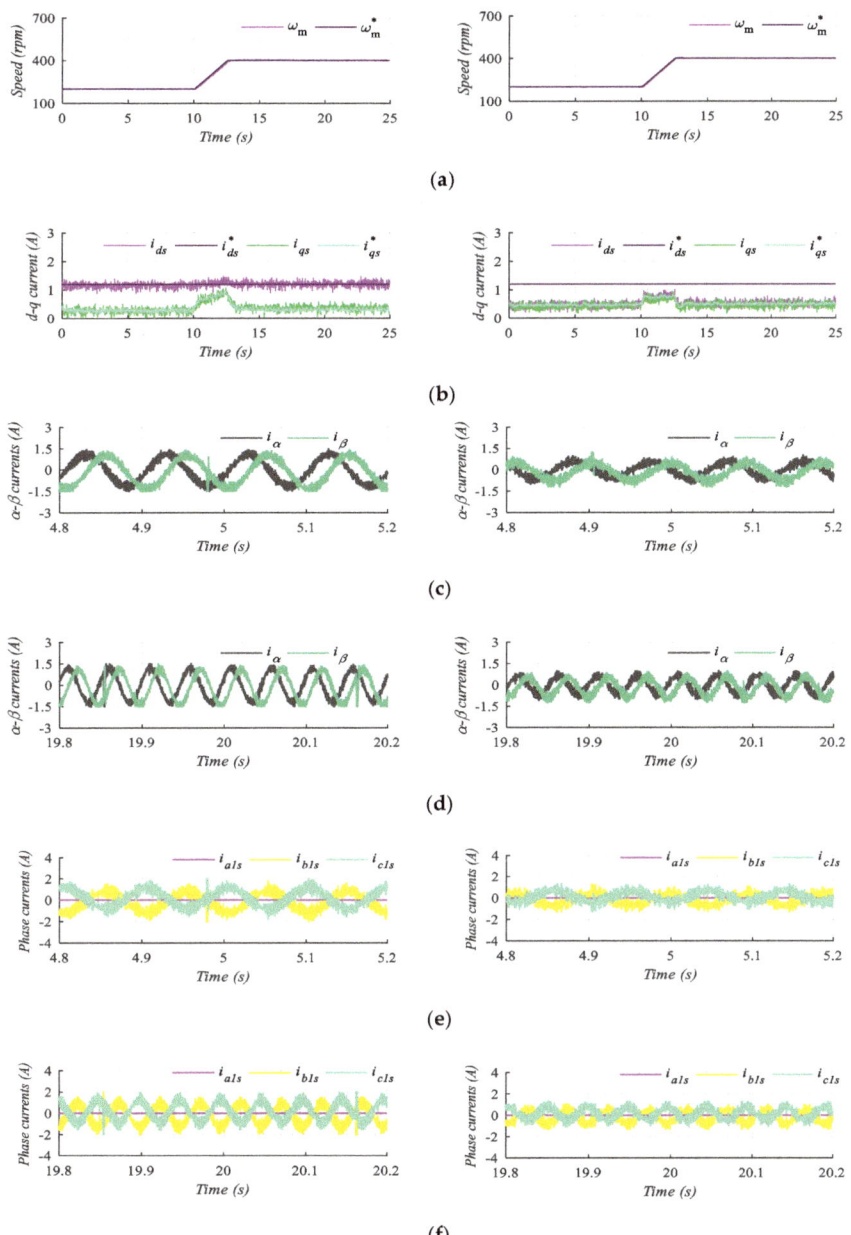

Figure 9. Post-fault dynamic response using VV-MPC technique (left column) and the proposed EVV-MPC method (right column). From top to bottom: (**a**) Motor speed; (**b**) d-q currents; (**c**) zoom 1 of α-β currents; (**d**) zoom 2 of α-β currents; (**e**) zoom 1 of set 1 of phase currents; and (**f**) zoom 2 of set 1 of phase currents.

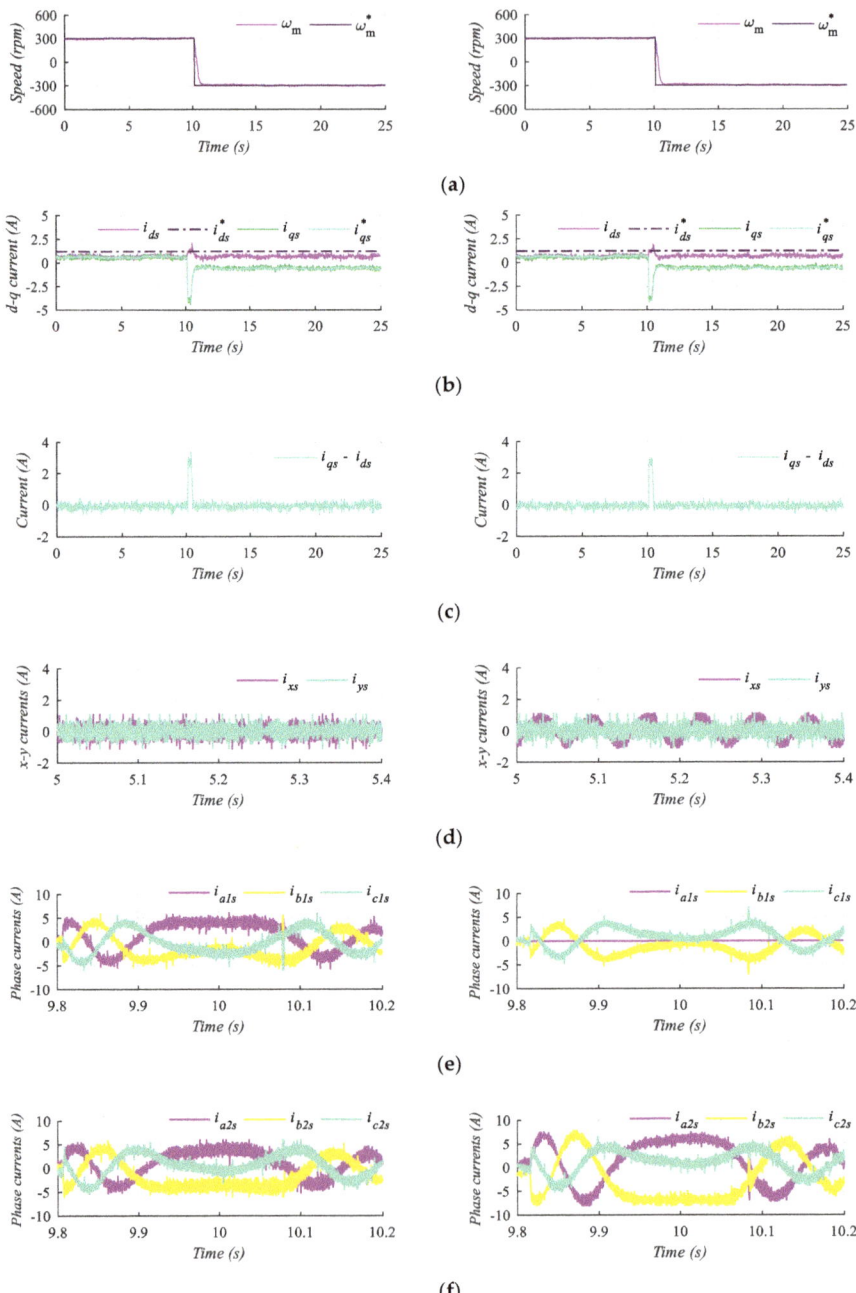

Figure 10. Reversal speed test in pre (left column) and post-fault situation (rigth column). From top to bottom: (**a**) Motor speed; (**b**) *d-q* currents; (**c**) difference between *q*- and *d*-currents; (**d**) zoom *x-y* currents; (**e**) zoom of set 1 of phase currents; and (**f**) zoom of set 2 of phase currents.

6. Conclusions

Efficiency and robustness are a must in industrial electric drives. Even though multiphase machines offer these two features, they are typically achieved at the expense of a high complexity. While the efficiency can be improved using either search control or loss model control, the fault-tolerant operation has been traditionally accomplished with a post-fault control reconfiguration. This work presents a simpler approach that reduces the system losses and increases the drive reliability. The approach is based on the MPC technique, the virtual voltage vector concept and a modified cost function that provides the capability to automatically operate in pre- and post-fault scenarios with excellent speed tracking and reduced copper losses. Key features of the proposal are the simplicity and universality. After a fault occurrence, the proposed strategy keeps on regulating the speed at optimum efficiency even when the fault has not even been detected, making the controller immune to fault detection delays and errors. Although the proposal has been experimentally tested in six-phase IM motor drives, it can be extended to multiphase drives with a higher number of phases if distributed windings are considered.

Author Contributions: Conceptualization and methodology: A.G.-P., I.G.-P., and M.J.D.; software: A.G.-P. and I.G.-P.; validation: M.J.D., I.G.-P., and F.B.; formal analysis and investigation: all authors; resources: all authors; data curation: all authors; writing—original draft preparation: all authors; writing—review and editing: all authors; visualization: all authors; supervision: I.G.-P., M.J.D., and F.B.; project administration: I.G.-P., M.J.D., and F.B.; funding acquisition: I.G.-P., M.J.D., and F.B.

Funding: This research was funded by the Spanish Government under the Plan Estatal 2017–2020 with the reference RTI2018-096151-B-I00.

Conflicts of Interest: The authors declare no conflict of interest.

References

1. Global Wind Energy Council (GWEC). Global Status of Wind Power. Available online: https://gwec.net/global-figures/wind-energy-global-status/ (accessed on 11 August 2019).
2. Levi, E.; Barrero, F.; Duran, M.J. Multiphase machines and drives-Revisited. *IEEE Trans. Ind. Electron.* **2016**, *63*, 429–432. [CrossRef]
3. Barrero, F.; Duran, M.J. Recent advances in the design, modeling and control of multiphase machines—Part 1. *IEEE Trans. Ind. Electron.* **2016**, *63*, 449–458. [CrossRef]
4. Duran, M.J.; Barrero, F. Recent advances in the design, modeling and control of multiphase machines—Part 2. *IEEE Trans. Ind. Electron.* **2016**, *63*, 459–468. [CrossRef]
5. Liserre, M.; Cardenas, R.; Mdinas, M.; Rodriguez, J. Overview of Multi-MW Wind Turbines and Wind Parks. *IEEE Trans. Ind. Electron.* **2011**, *58*, 1081–1095. [CrossRef]
6. Gamesa Technological Corporation. Gamesa 5.0 MW. Available online: http://www.gamesacorp.com/recursos/doc/productos-servicios/aerogeneradores/catalogo-g10x-45mw.pdf (accessed on 8 August 2019).
7. Gonzalez-Prieto, I.; Duran, M.J.; Rios-Garcia, N.; Barrero, F.; Martin, C. Open-Switch Fault Detection in Five-Phase Induction Motor Drives Using Model Predictive Control. *IEEE Trans. Ind. Electron.* **2018**, *65*, 3045–3055. [CrossRef]
8. Munim, W.N.W.A.; Duran, M.J.; Che, H.S.; Bermudez, M.; Gonzalez-Prieto, I.; Rahim, N.A. A Unified Analysis of the Fault-Tolerance Capability in a Six-Phase Induction Motor Drives. *IEEE Trans. Power Eelctron.* **2017**, *32*, 7824–7836. [CrossRef]
9. Gonzalez-Prieto, I.; Duran, M.J.; Bermudez, M.; Barrero, F.; Martin, C. Assessment of Virtual-Voltage-based Model Predictive Controllers in Six-Phase Drives under Open-Phase Fault. *IEEE J. Emerg. Sel. Top. Power Electron.* **2019**. [CrossRef]
10. Che, H.S.; Duran, M.J.; Levi, E.; Jones, M.; Hew, W.; Rahim, N.A. Post fault Operation of an Asymmetrical Six-Phase Induction Machine with Single and Two Isolated Neutral Points. *IEEE Trans. Power Electron.* **2014**, *29*, 5406–5416. [CrossRef]
11. Kirschen, D.S.; Novotny, D.W.; Lipo, T.A. Optimal efficiency control of an induction motor drive. *IEEE Trans. Energy Convers.* **2012**, *27*, 958–967.

12. Mesemanolis, A.; Mademlis, C.; Kioskeridis, I. High efficiency control for a wind energy conversion system with induction generator. *IEEE Trans. Energy* **2012**, *27*, 958–967. [CrossRef]
13. Takahashi, I.; Noguchi, T. A new quick-response and high efficiency control strategy of an induction motor. *IEEE Trans. Ind. Appl.* **2006**, *22*, 820–827. [CrossRef]
14. Lorenz, R.D.; Yang, S.M. Efficiency-optimized flux trajectories for closed cycle operation of field-orientation induction machine drives. *IEEE Trans. Ind. Appl.* **1992**, *28*, 574–580. [CrossRef]
15. Sul, S.K.; Park, M.H. A novel technique for optimal efficiency control of a current-source inverter-fed induction motor. *IEEE Trans. Power Electron.* **1988**, *3*, 192–199. [CrossRef]
16. Gonzalez-Prieto, I.; Duran, M.J.; Barrero, F.; Bermudez, M.; Guzman, H. Impact of post-fault flux adaptation on six-phase induction motor drives with parallel converters. *IEEE Trans. Power Electron.* **2017**, *32*, 515–528. [CrossRef]
17. Gonzalez-Prieto, A.; Gonzalez-Prieto, I.; Duran, M.J. Efficient Predictive Control with Natural-Fault Tolerance for Multiphase Induction Machines. In Proceedings of the 45th Annual Conference of the IEEE Industrial Electronics Society (IECON2019), Lisbon, Portugal, 14–17 October 2019.
18. Zhao, Y.; Lipo, T.A. Space Vector PWM control of dual three-phase induction machine using vector space decomposition. *IEEE Trans. Ind. Appl.* **1995**, *31*, 1100–1109. [CrossRef]
19. Duran, M.J.; Levi, E.; Barrero, F. Multiphase Electric Drives: Introduction. In *Wiley Encyclopedia of Electrical and Electronics Engineering*; Webster, J.G., Ed.; Wiley: Hoboken, NJ, USA, 2017; pp. 1–26.
20. González-Prieto, I.; Duran, M.J.; Aciego, J.J.; Martin, C.; Barrero, F. Model Predictive Control of Six-Phase Induction Motor Drives Using Virtual Voltage Vectors. *IEEE Trans. Ind. Electron.* **2018**, *65*, 27–37. [CrossRef]
21. Aciego, J.J.; Gonzalez-Prieto, I.; Duran, M.J. Model predictive Control of Six-Phase Induction Motor Drives Using Two Virtual Vectors. *IEEE J. Emerg. Sel. Top. Power Electron.* **2019**, *7*, 321–330. [CrossRef]
22. Tani, A.; Mengoni, M.; Zarri, L.; Serra, G.; Casadei, D. Control of Multiphase induction motors with and odd number of phases under open-circuit phase faults. *IEEE Trans. Power Electron.* **2012**, *27*, 565–577. [CrossRef]
23. Duran, M.J.; Gonzalez-Prieto, I.; Barrero, F.; Levi, E.; Zarri, L.; Mengoni, M. A simple Braking Method for Six-Phase Induction Motor Drives with Unidirectional Power Flow in the Base-Speed Region. *IEEE Trans. Ind. Electron.* **2017**, *64*, 6032–6041. [CrossRef]
24. Bermudez, M.; Gonzalez-Prieto, I.; Barrero, F.; Guzman, H.; Kestelyn, X.; Duran, M.J. An Experimental Assessment of Open-Phase Fault-Tolerant Virtual-Vector-Based Direct Torque control in Five-Phase Induction Motor Drives. *IEEE Trans. Power Electron.* **2018**, *33*, 2774–2784. [CrossRef]
25. Duran, M.J.; Gonzalez-Prieto, I.; Rios-Garcia, N.; Barrero, F. A Simple fast and Robust Open-Phase Fault Detection Technique for Six-Phase Induction Motor Drives. *IEEE Trans. Power Electron.* **2018**, *33*, 547–557. [CrossRef]
26. Yepes, A.G.; Riveros, J.A.; Candoy, J.D.; Barrero, F.; Lopez, Ó.; Bogado, B.; Jones, M.; Levi, E. Parameters identification of multiphase induction machines with distributed windings-Part 1: Sinusoidal excitation methods. *IEEE Trans. Energy Conv.* **2012**, *27*, 1056–1066. [CrossRef]
27. Riveros, J.A.; Yepes, A.G.; Barrero, F.; Doval-Gandoy, J.; Bogado, B.; Lopez, O.; Jones, M.; Levi, E. Parameters identification of multiphase induction machines with distributed windings-Part 2: Time domain techniques. *IEEE Trans. Energy Convers.* **2012**, *27*, 1067–1077. [CrossRef]

© 2019 by the authors. Licensee MDPI, Basel, Switzerland. This article is an open access article distributed under the terms and conditions of the Creative Commons Attribution (CC BY) license (http://creativecommons.org/licenses/by/4.0/).

Article

Model-Based Predictive Current Controllers in Multiphase Drives Dealing with Natural Reduction of Harmonic Distortion

Cristina Martin [1], Federico Barrero [1,*], Manuel R. Arahal [2] and Mario J. Duran [3]

1. Electronic Engineering Department, University of Seville, 41092 Seville, Spain; martintorrescristina@gmail.com
2. System and Automatic Engineering Department, University of Seville, 41092 Seville, Spain; arahal@us.es
3. Electrical Engineering Department, University of Malaga, 29071 Malaga, Spain; mjduran@uma.es
* Correspondence: fbarrero@us.es; Tel.: +34-954-48-13-04

Received: 14 March 2019; Accepted: 29 April 2019; Published: 3 May 2019

Abstract: An important drawback in the application of model-based predictive controllers for multiphase systems is the relatively high harmonic content. Harmonics arise due to the fixed sampling-time nature and the absence of modulation methods in the control technique. Recent research works have proposed different procedures to overcome this disadvantage at the expense of increasing the complexity of the controller and, in most cases, the computational requirements. There are, however, natural ways to face this harmonic generation that have been barely explored in the scientific literature. These alternatives include the use of variable sampling times or the application of the observer theory, whose utility has been stated without excessively increasing the computational cost of the controller. This paper presents the basis of both methodologies, analyzing their interest as natural alternatives to mitigate the generation of harmonic components in modern electrical drives when using predictive controllers. A five-phase induction machine is used as a case example to experimentally validate the study and draw conclusions.

Keywords: predictive current control; harmonic distortion; multiphase drives; observer; variable sampling

1. Introduction

The increasing interest for multiphase drives in real applications [1,2], added to the complexity of designing appropriate controllers for these multivariable systems, have put the emphasis on model predictive control methods (MPC) and particularly on the finite control set MPC (FCS-MPC) [3]. The FCS-MPC is a kind of fast direct control method that commands the power converter without using pulse width modulation (PWM) blocks, providing excellent transient performance and lower switching frequency than PWM blocks with conventional proportional-integral controllers (PI-PWM), under comparable conditions [4,5]. This issue has been extensively investigated in [6], where FCS-MPC and PI-PWM current controllers are compared, concluding that the FCS-MPC provides a faster transient evolution at the expense of a lower steady-state performance, something that is, in general, inevitable in multiphase drives due to the existence of nonflux/torque producing current components. Additionally, the simple and multi-objective formulation of the FCS-MPC algorithm makes it an excellent option in multiphase drives, being that five-phase induction machine (IM) is one of the most investigated configurations [7].

However, an important drawback appears in the FCS-MPC implementation, which is the high current/voltage harmonic content. This problem has been recently examined in [8], concluding that the fixed-time discretization nature of the control method, along with the fact that only one of the possible power converter states is applied during each sampling interval, favour the appearance of not only high magnitude harmonics but also inter-harmonics and electrical noise. Some recent solutions based on the selective harmonic elimination concept can reduce harmonics of the integer multiples of the fundamental frequency [9,10], but they do not cancel inter-harmonics and electrical noise.

A careful design of the cost function, which represents the control objectives of the FCS-MPC, can also help in the reduction of the harmonic content [11]. For example, a precise tuning of the weighting factors that weight each control objective can be decisive [4,6,12], as well as the limitation of the commutation frequency in the converter by the restriction of the available changes in the switches of the converter's legs [13]. However, these techniques generally increase the controller complexity and the computational requirements, another important handicap in the application of FCS-MPC methods to multiphase drives. Furthermore, they can lead to suboptimal solutions when not all the possible control actions are taken into account. In addition, there exists an interdependence between the harmonic content and other control aspects, such as the switching frequency or the operating point, which can be seen as fundamental trade-offs that the cost function design cannot completely bypass [14].

A quite different alternative for the harmonic mitigation consists in adding a modulation stage in the FCS-MPC algorithm [15] or applying more than one switching state of the power converter during the same sampling time [16], which is in essence a kind of modulation. However, these techniques produce higher switching frequencies than conventional FCS-MPC methods when identical sampling time is imposed, and could increase the high computational cost of the predictive controller.

The use of simpler natural solutions can alleviate the harmonic problem that the previously cited techniques suffer. One of them is the newly proposed variable sampling time lead pursuit controller (VSTLPC) [17], which introduces the concept of non-uniform sampling time as a new degree of freedom in the model-based predictive technique. In this way, both the switching state of the power converter and its time of application are optimally selected between all the possibilities without the necessity of a cost function and with an affordable computational cost. A different alternative consists of the improvement of the predictive model, since the selected control action depends on it. In this context, the observer theory has been recently incorporated in the FCS-MPC for the estimation of non-measurable parts of the system model, leading to significant improvement of the system performance. Rotor current observers based on the Luenberger theory and Kalman filters are usually applied in the FCS-MPC current control of multiphase IM replacing the traditional backtracking procedures [18,19]. This work focuses on the study of VSTLPC and rotor-current observers as natural ways to reduce the harmonic distortion and electrical noise in predictive controllers. The basis of compared techniques will be reviewed in Sections 2 and 3, where a five-phase IM drive is used as a well-known case example of multiphase drives. Experimental results to corroborate the utility of these techniques are presented in Section 4, while the obtained conclusions are summarized in the last section.

2. Rotor Estimation in FCS-MPC Techniques: The Observer Approach

Considering a five-phase IM drive supplied by a two-level five-phase voltage source inverter (VSI) as the controlled system under study, the general scheme of the applied FCS-MPC current control is illustrated in Figure 1a. The main goal is to find the switching state (S_{opt}) that forces the stator currents (i_s) to follow the references (i_s^*). To this end, a prediction of the future stator currents (i_s^p) is computed using an electrical model of the IM drive (predictive model) and the measured i_s and rotor speed (ω_r). The prediction and references are then compared inside a predefined cost function (J) to find the switching state that minimizes their difference. The algorithm is iterated and repeated using a constant sampling period.

In this process, the predictive model plays an important role and the best agreement with the real system will improve the predictions and, consequently, the performance of the regulated system.

The five-phase IM can be represented, using the well-known vector space decomposition approach, by a set of equations expressed in the two orthogonal α–β and x–y subspaces as follows:

$$\begin{aligned} \dot{x}(t) &= f(x(t), S(t)) \\ x_s(t) &= C\, x(t), \end{aligned} \quad (1)$$

where the state variables are the stator and rotor currents $x = (i_{s\alpha}, i_{s\beta}, i_{sx}, i_{sy}, i_{r\alpha}, i_{r\beta})$, the control signal is the switching state of the VSI that is arranged in vector $S = (S_A, S_B, S_C, S_D, S_E) \in \mathbb{B}^5$ with $\mathbb{B} = \{0, 1\}$, the output signals are the stator currents $x_s = (i_{s\alpha}, i_{s\beta}, i_{sx}, i_{sy})$, and function f depends on the IM parameters, the spatial distribution of the windings, the VSI connections and the instant value of the rotor speed. Further details of the multiphase IM drive modeling can be encountered in [20], and in [18] for the particular five-phase case. The discretization of these non-linear equations provides the predictive model (2), normally using the forward Euler method or a more complicated technique based on the Cayley–Hamilton theorem, which improves the tracking and prediction performance [21].

$$\begin{aligned} x^p(k+1) &= x(k) + T_s f(x(k), S(k)) \\ x_s^p(k+1) &= C\, x^p(k+1), \end{aligned} \quad (2)$$

In any case, a second-step prediction $x^p(k+2)$ is usually applied to compensate the delay that introduces the computation of the control algorithm [4]. Then, the cost function, usually defined as in (3) from the squared error between the predictions and reference currents $\hat{e} = i_s^*(k+2) - i_s^p(k+2)$, is computed for all the available switching vectors of the VSI to obtain the next control action to be applied.

$$J = \|\hat{e}_{\alpha\beta}\|^2 + \lambda_{xy}\|\hat{e}_{xy}\|^2. \quad (3)$$

This cost function includes a weighting factor λ_{xy} to put more or less emphasis in the x–y control plane, which is related to the copper losses in our case since sinusoidal winding distribution is assumed in the IM. The tuning of this parameter is not a simple issue [11], but a value of 0.5 is usually accepted because it provides a good trade-off between both planes [6]. Stator current references in the d–q rotating reference frame are imposed and then rotated using the inverse of the Park transformation D^{-1} and the rotational angle θ [6], obtaining α–β current references. Furthermore, x–y references are set to zero to minimize the stator copper losses.

While stator currents are measured, rotor ones are commonly estimated using a simple backtracking procedure that consists in lumping into term G all non-measurable quantities and other uncertainties of the system. This term is recalculated every sampling period using the system model and past values of the measured variables. Thus, the predictive model can be rewritten as:

$$x_s^p(k+1) = x_s(k) + T_s f_s(x_s(k), S(k)) + G^e(k), \quad (4)$$

being f_s the part of the function f in (2) related only to the stator currents, and superscript e stands for estimated values. Using this method, the rotor estimation error will be compensated at each sampling period, being this effect accentuated by smaller sampling periods. However, even a small amount of electrical noise has an important effect in the prediction error, which can even lead to a wrong selection of the switching vector and produce a high disturbance in the tracking performance. Another commonly used backtracking procedure is the one applied in [6], where an open-loop observer based on the system model is used to obtain estimated values of the rotor variables as follows:

$$x_r^e(k) = x_r^e(k-1) + T_s f_r(x(k-1), S(k-1)). \quad (5)$$

Rotor currents are updated every k instant using the previous values of the measured variables and the applied switching state. Notice that function f_r is the part of function f in (2) that provides the rotor current values. Although a more precise rotor current estimation can be obtained with this approximation, the previous problems still remain and the noise can degrade the control performance.

An alternative to aforementioned techniques goes through the use of closed-loop observers, where the rotor current estimation is done using Kalman filters or Luenberger-based observers. Among them, the full-order version of the Luenberger observer has shown the best rotor estimation result at the expense of a slight increment in the computational cost [18]. In this Luenberguer-based approach, estimation of both stator and rotor currents x^e is computed using the system model (1) plus a correction term weighted by the Luenberger matrix L:

$$\dot{x}^e(t) = f(x^e(t), S(t)) - L(C\,x^e(t) - x_s(t)). \qquad (6)$$

The design of the observer consists in a pole placement problem in which matrix L is obtained as a result. A good practice consists of placing the observer's eigenvalues in the position defined by the roots of a Butterworth filter polynomial, permitting a fast convergence towards zero of the estimation error, as well as a well-dumped dynamic without compromising the stability. Although the design of the observer requires the solution of this problem, it can be done off-line and simple expressions of L can be obtained for all the operating speed range that, in turn, does not excessively increase the computational cost of the controller. Also, the Luenberger observer has demonstrated to be more robust under model uncertainties than previous backtracking procedures, showing better rotor current estimations and, consequently, improving the performance of the controlled system.

3. Variable Sampling Time in Predictive Controllers

An alternative model-based predictive current controller named VSTLPC is detailed in [17], where the sampling time is a new degree of freedom that is calculated by the control algorithm at each iteration. The schematic representation of the VSTLPC current control applied to a five-phase IM is detailed in Figure 1b. Similarly to FCS-MPC, the optimal switching state (S_{opt}) of the power converter is selected in order to produce the desired stator current response defined by the reference (i_s^*). However, the application time of the converter state is not fixed and equal to the sampling time, but it is also decided by the controller.

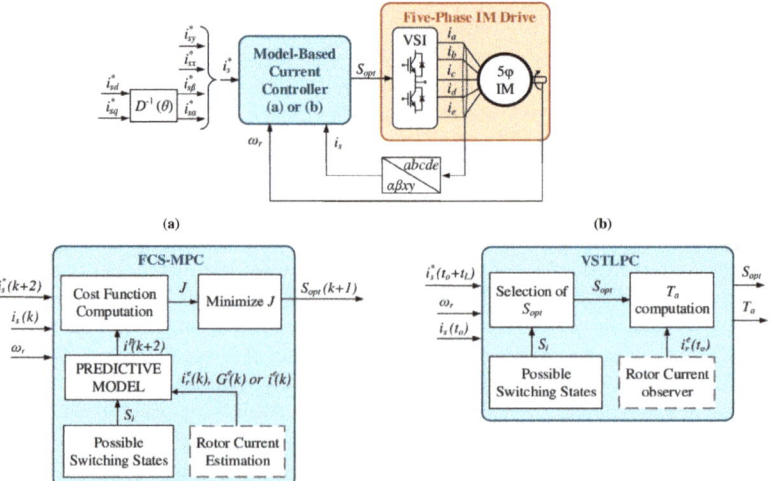

Figure 1. Simplified scheme of the five-phase induction machine (IM) drive current controller using (a) finite control set (FCS)-predictice control methods (MPC) and (b) variable sampling time lead pursuit controller (VSTLPC) techniques.

The control algorithm starts with the selection of S_{opt} based on the measurement of the stator currents (i_s) and rotor speed (ω_r) and using the lead-pursuit concept: hitting a moving target requires some anticipation, since it takes some time for the control action to produce an effect on the system and during such time the target changes its position. In this way, the controller points to an advanced stator current reference $i_s^*(t_o + t_L)$, where t_o is the present time instant and t_L is the anticipation time or lead time. Then, the switching vector that produces the closest trajectory of stator currents to the reference is selected. The ideal trajectory would be the one formed by measured currents and advanced references, which is defined by $(x_s^*(t_o + t_L) - x_s(t_o))$ in our case. Following that $f_s(x, S)$ is a vector that determines how stator currents evolve, the cosine of the angle between vectors $f_s(x(t_o), S)$ and $(x_s^*(t_o + t_L) - x_s(t_o))$ is the maximum for the switching state that minimizes the deviation from the objective. Consequently, the optimal switching vector S_{opt} is selected through the definition of the scalar product:

$$S_{opt} = \underset{S_i}{\mathrm{argmax}}\ \frac{(x_s^*(t_o + t_L) - x_s(t_o)) \cdot f_s(x(t_o), S_i)}{\|x_s^*(t_o + t_L) - x_s(t_o)\|\, \|f_s(x(t_o), S_i)\|}. \tag{7}$$

The above expression is an optimization problem that takes into account all possible switching states. It is necessary to remember that vector $x(t_o)$ in function f_s is formed by measured stator currents and the estimated rotor ones. Note that rotor currents are obtained in [17] using the Luenberger observer detailed in the previous section.

The application time T_a of the selected voltage vector is obtained minimizing the deviation between the stator references and predicted currents:

$$T_a = \underset{T}{\mathrm{argmin}}\ \left\|x_s^*(t_o + t_L) - x_s^p(t_o + T)\right\|, \tag{8}$$

where $x_s^p(t_o + T)$ is obtained using the system Equation (2) for the selected S_{opt}. This minimization problem is finally solved using:

$$T_a = (x_s^*(t_o + t_L) - x_s(t_o))^\top \frac{f_s(x(t_o), S_{opt})}{\|f_s(x(t_o), S_{opt})\|^2}. \tag{9}$$

After that, a receding horizon process is applied where the selected vector is released during the obtained application time and the control algorithm is repeated. Comparing with FCS-MPC techniques, the VSTLPC method permits a fine resolution of commuting times thanks to the non-uniform sampling, which can mitigate the generated harmonic distortion. This hypothesis will be analyzed in the next section, where a comparative analysis of the generated harmonic distortion using FCS-MPC and VSTLPC techniques is done.

4. Harmonic Distortion Using FCS-MPC and VSTLPC Techniques: Comparative Analysis

A current control performance analysis of the revised controllers is done using the experimental test bench shown in Figure 2. The main component is a 30-slot symmetrical five-phase IM with distributed windings, whose electrical parameters are gathered in Table 1. These have been obtained through the experimental tests described in [22,23]. Two three-phase two-level inverters from Semikron (SKS22F modules) supply the IM, and an external DC-link voltage of 300 V is connected to them. The multiphase system is controlled using a MSK28335 Technosoft board that includes a TMS320F28335 digital signal processor (DSP). The rotor mechanical speed (ω_m) is measured using a GHM510296R/2500 digital encoder. Finally, an independently controlled DC machine is used to impose an external variable load torque in the shaft of the IM.

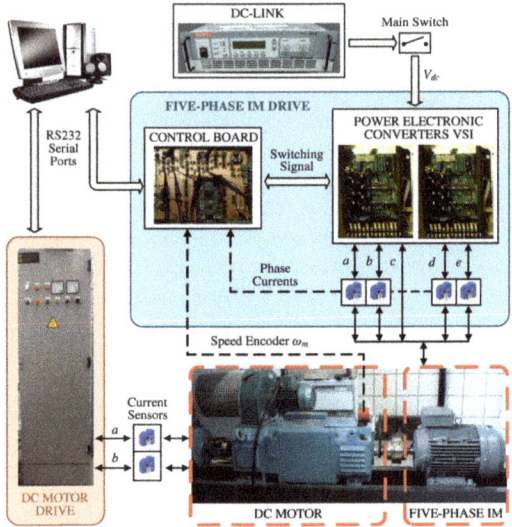

Figure 2. Experimental test rig.

Table 1. Estimated parameters of the IM.

Parameter		Value
Stator resistance	R_s (Ω)	19.45
Rotor resistance	R_r (Ω)	6.77
Stator leakage inductance	L_{ls} (mH)	100.7
Rotor leakage inductance	L_{lr} (mH)	38.6
Mutual inductance	L_m (mH)	656.5
Mechanical nominal speed	ω_n (rpm)	1000
Nominal torque	T_n (N·m)	4.7
Nominal current	I_n (A)	2.5
Pole pairs	P	3

The controllers used in the comparison are the FCS-MPC technique with the conventional backtracking procedure (MPC-C1) and the open-loop observer (MPC-C2), the FCS-MPC method with a closed-loop rotor current observer (MPC-OB), and the VSTLPC. Equal cost functions are applied in MPC-C1, MPC-C2 and MPC-OB with a weighting factor of 0.5, for the reasons presented in Section 2. The Luenberger rotor current observer is designed using a fourth order Butterworth filter (10), since the system presents two real poles that are maintained in the design of the observer:

$$B_4(s) = T_B^4 s^4 + 2.61 T_B^3 s^3 + 3.41 T_B^2 s^2 + 2.61 T_B s + 1. \tag{10}$$

A value of $T_B = 0.001$ s has been optimally selected by simulations in order to produce the lowest observation error in all speed range. Regarding the sampling time, it is imposed as $T_s = 100$ μs for the three FCS-MPC techniques with fixed discretization. For the case of the VSTLPC method, the sampling time is limited by a minimum value of $T_{min} = 100$ μs to make a fair comparison with the other controllers, and a maximum value of $T_{max} = 300$ μs to avoid larger sampling periods that could deteriorate the control performance [17]. The lead time is set to $t_L = 100$ μs.

First, several steady-state tests have been carried out for each controller and the performance analysis is done on the basis of the mean square tracking error of the phase currents (RMSe$_p$), the total harmonic distortion in the phase currents (THD$_p$), and the number of commutations per cycle (NCPC) in the VSI legs. In all tests, the d-current reference is fixed to 0.57 A to produce the rated flux and the system is closed-loop speed controlled using an outer PI regulator that provides the q-current reference. In this way, it is possible to drive the machine to a constant rotor speed in the range of 50 rpm to 700 rpm. In addition, a variable load torque between the 40% to the 70% of the nominal torque is imposed. The obtained results are plotted in Figure 3. In each column, the VSTLPC technique is compared with one of the other controllers in terms of the three aforementioned figures of merits. In such a way, the interest of including the non-fixed sampling against the FCS-MPC methods is revealed.

Regarding the current tracking performance and the harmonic content, lower values of $RMSe_p$ and THD_p are observed in the VSTLPC technique in all considered operating conditions when it is compared with the MPC-C1 and MPC-C2 methods, being the difference bigger in the first comparison. However, the opposite occurs when the VSTLPC and MPC-OB techniques are compared. In this case, the $RMSe_p$ and the THD_p values are lower for the MPC-OB, indicating that the inclusion of the rotor current observer in the FCS-MPC is enough to produce a significant improvement in the current control performance with respect to the conventional techniques. It must be noticed that the backtracking procedure based on the open-loop rotor current observer (MPC-C2) provides better results than the most conventional rotor estimation approach (MPC-C1), demonstrating a higher robustness to external disturbances as it was stated in Section 2. In terms of the number of commutations per cycle, the VSTLPC produces the highest values, being this effect accentuated by the decrease of the speed, while the MPC-OB presents the lowest values in most cases.

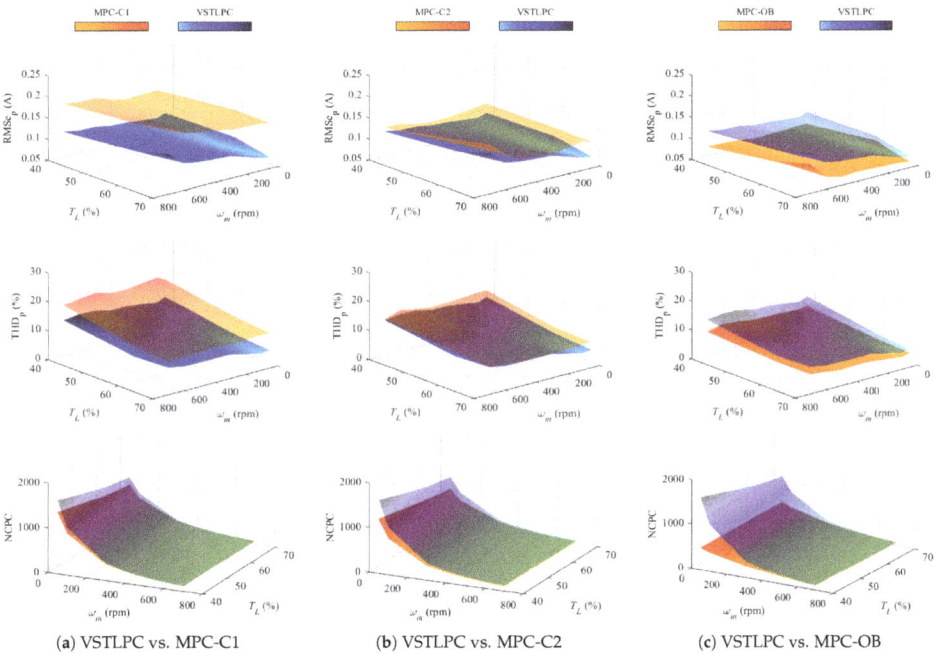

(a) VSTLPC vs. MPC-C1   (b) VSTLPC vs. MPC-C2   (c) VSTLPC vs. MPC-OB

Figure 3. Experimental root mean square error of phase currents ($RMSe_p$), total harmonic distortion (THD) and number of commutations per cycle (NCPC) values for each controller under different operating conditions.

It is interesting to mention that, in general, the evolution with the speed and the torque of all performance parameters is similar for all considered control techniques, regardless of the different values between them.

To complete the previous analysis, Figures 4 and 5 show the current control performance of all considered controllers for two of the analyzed operating conditions: 100 rpm and 60% of the nominal torque (Figure 4), and 600 rpm and 70% of the nominal torque (Figure 5). For the first experiment, the circular α–β and x–y current trajectories and their references appear in the upper plots, while in the second test the evolution with the time of the α and x currents are shown. In both tests, the spectrum of the a–phase current is plotted and zoomed in the lower plots. These spectrums show harmonics and inter-harmonics that have been measured following the guidelines of the ICE standard [24], but after adapting the normative to the case under study. Thus, nine and 10 cycles of the current signal have been used for the spectrum calculation in the 100 rpm and 600 rpm cases, respectively, in order to guarantee a good resolution.

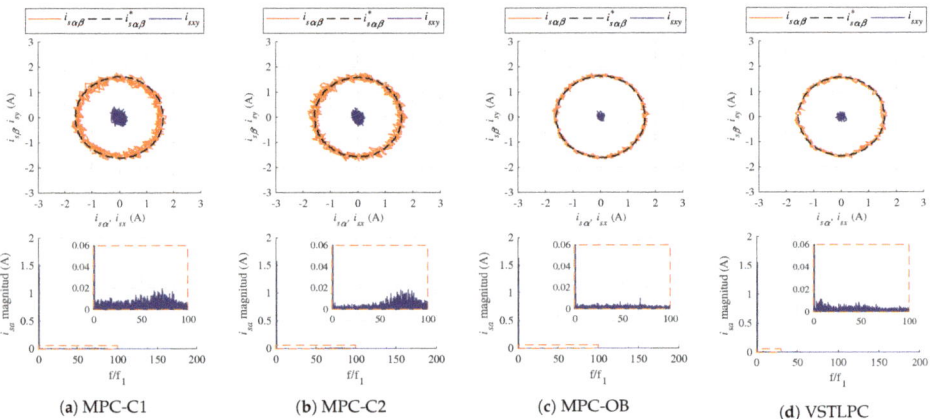

Figure 4. Experimental steady-state test for 100 rpm and a load torque of 60% of the nominal one. Upper plots present the α–β and x–y current trajectories, and the lower plots present the a-current spectrum.

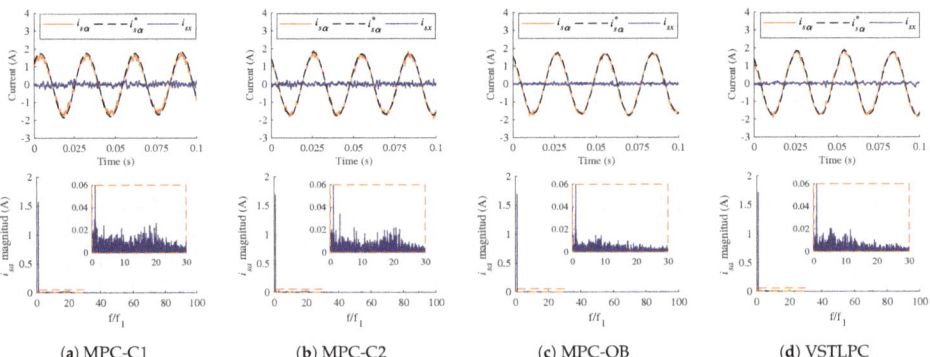

Figure 5. Experimental steady-state test for 600 rpm and a load torque of 70% of the nominal one. Upper plots present the α and x current trajectories, and the lower plots present the a-current spectrum.

It can be seen that the worst current tracking performance is obtained with the MPC-C1 technique, which presents a small offset in the tracking of the α–β currents. This offset is a characteristic of most predictive controllers [3] but it is significantly reduced by the application of the closed-loop observer and the variable sampling. The harmonic and noise content is also reduced with the new controllers (MPC-OB and VSTLPC), as evidenced by the lower ripple in the α–β–x–y currents and the reduced magnitude in the current spectrum in comparison with conventional FCS-MPC techniques (MPC-C1 and MPC-C2). This, in turn, leads to a more efficient flux and torque production with lower copper losses. Focusing on the current spectrum, it is interesting to see that the MPC-C1 technique presents a more continuous spectrum with high magnitude of harmonics in a large frequency domain, while the MPC-C2 method shows significant harmonic distortion principally at high frequencies (this effect is more accentuated at lower speeds and loads). Conversely, the MPC-OB and VSTLPC approaches effectively reduce the harmonic magnitude in all the frequency domain. Although the VSTLPC presents some peaks of distortion at low frequencies for specific operating points, the total harmonic distortion is still lower than in conventional FCS-MPC methods (Figure 3).

Three dynamic tests have also been done in order to validate the current control performance during the transient. The first one consists in a speed reversal test from −500 rpm to 500 rpm imposing a load torque equal to the 60% of the nominal one. The second test is a speed step from 0 rpm to 500 rpm imposing a load torque of the 60% too. Finally, the third test is a torque step from 0% to 60% of the nominal torque at 500 rpm. Since all controllers present similar speed response in each test, only the speed evolution for the case of the VSTLPC method is presented in Figure 6 for simplicity reasons. Diversely, the d–q currents evolution with time for each controller is presented in Figures 7–9 for the speed reversal, the speed step and the torque step experiments, respectively. Regarding the transient performance, it can be stated that it is quite similar for all controllers. This fact demonstrates that the inclusion of the closed-loop rotor current observer and the variable sampling time does not deteriorate the fast transient performance that characterizes the predictive controller. Furthermore, superior current tracking and lower harmonic distortion are provided by the MPC-OB and VSTLPC techniques, as it was expected by the previous steady-state results. This is evidenced by the $d - q$ currents performance in Figures 7–9, where the current ripple and the previously cited offset are higher when using the conventional MPC-C1 and MPC-C2 methods even during the transient. Consequently, the harmonic content is also higher in that cases comparing to the recently proposed current control approaches.

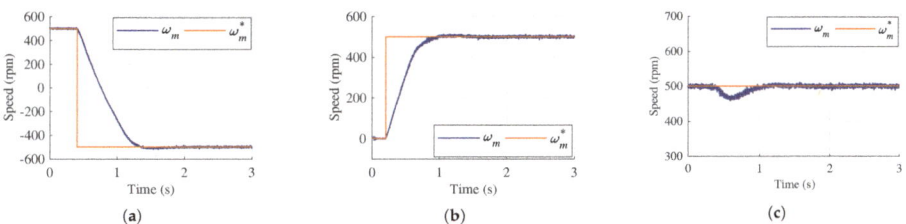

Figure 6. Rotor speed dynamic for the VSTLPC: (**a**) reversal test from −500 rpm to 500 rpm, (**b**) speed step test from 0 rpm to 500 rpm, both tests with a load torque of 60%, and (**c**) torque step test from 0% to 60% of the nominal torque at 500 rpm.

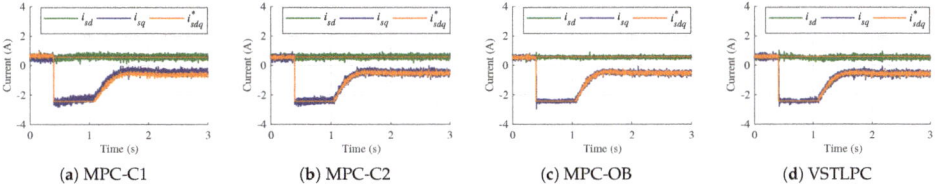

Figure 7. Evolution of d–q currents and their references for each controller in a reversal test from −500 rpm to 500 rpm with a load torque of the 60%.

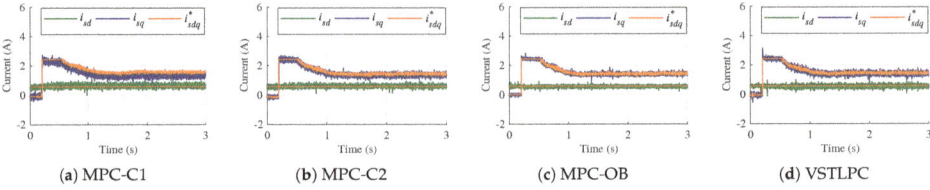

Figure 8. Evolution of *d–q* currents and their references for each controller in a speed step test from 0 rpm to 500 rpm with a load torque of the 60%.

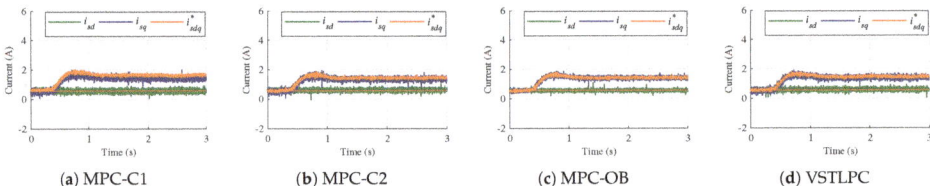

Figure 9. Evolution of *d–q* currents and their references for each controller in a torque step test from 0% to 60% of the nominal torque at 500 rpm.

To conclude the comparative assessment, the computational cost of analyzed controllers was studied. The conventional MPC-C1 and MPC-C2 approaches require, in the DSP-based experimental system, a computational cost around 32 µs and 34 µs, respectively. On the other hand, the MPC-OB and the VSTLPC techniques require 36 µs and 55 µs, respectively. This increment is completely affordable taking into account that the minimum sampling time is 100 µs. It must be noticed that the computational burden of the VSTLPC strongly depends on the operating point, as it was stated in [17], being the previously indicated computational cost a mean value.

To summarize, the VSTLPC and MPC-OB techniques outperform the conventional FCS-MPC methods in terms of harmonic content and tracking performance, but the closed-loop observer provides the best results in all the operating range. Regarding the NCPC, the VSTLPC provides the highest values while the lowest values are produced, in the most cases, by the MPC-OB. Note that the obtained benefits do not excessively increase the computational cost of the controller and do not compromise the fast transient response of the regulated system.

5. Conclusions

Model-based current predictive controllers applied to multiphase machines directly commands the power converter providing faster transient performance and lower switching frequencies that conventional PI-PWM methods. However, they suffer from high harmonic content in the electric variables, principally due to the inaccuracy of the prediction model and the fixed time discretization. In this work, a state of the art analysis of the situation has been done, where different predictive control techniques are compared, natural ways to reduce the obtained harmonic distortion are presented, and experiments are carried out using a five-phase IM drive as a case study is presented to conclude the benefits and drawbacks of the analyzed control methods.

Author Contributions: Conceptualization, F.B. and M.R.A.; methodology, C.M. and M.R.A; software, C.M.; validation, C.M.; formal analysis, C.M.; investigation, C.M.; resources, F.B.; data curation, C.M.; writing—original draft preparation, C.M. and F.B.; writing—review and editing, C.M., F.B., M.R.A. and M.J.D.; visualization, F.B.; supervision, F.B. and M.R.A.; project administration, C.M. and F.B.; funding acquisition, F.B.

Funding: This research received no external funding.

Acknowledgments: The authors would like to thank the University of Seville for its support under the V Research Plan, Action II.2.

Conflicts of Interest: The authors declare no conflict of interest.

References

1. Duran, M.J.; Levi, E.; Barrero, F. Multiphase Electric Drives: Introduction. In *Wiley Encyclopedia of Electrical and Electronics Engineering*; John Wiley & Sons, Inc.: Hoboken, NJ, USA, 2017; pp. 1–26.
2. Bojoi, R.; Rubino, S.; Tenconi, A.; Vaschetto, S. Multiphase electrical machines and drives: A viable solution for energy generation and transportation electrification. In Proceedings of the 2016 International Conference and Exposition on Electrical and Power Engineering (EPE), Iasi, Romania, 20–22 October 2016; pp. 632–639.
3. Kouro, S.; Perez, M.A.; Rodriguez, J.; Llor, A.M.; Young, H.A. Model Predictive Control: MPC's Role in the Evolution of Power Electronics. *IEEE Ind. Electron. Mag.* **2015**, *9*, 8–21. [CrossRef]
4. Arahal, M.; Barrero, F.; Toral, S.; Duran, M.; Gregor, R. Multi-phase current control using finite-state model-predictive control. *Control Eng. Pract.* **2009**, *17*, 579–587. [CrossRef]
5. Young, H.A.; Perez, M.A.; Rodriguez, J.; Abu-Rub, H. Assessing Finite-Control-Set Model Predictive Control: A Comparison with a Linear Current Controller in Two-Level Voltage Source Inverters. *IEEE Ind. Electron. Mag.* **2014**, *8*, 44–52. [CrossRef]
6. Lim, C.S.; Levi, E.; Jones, M.; Rahim, N.A.; Hew, W.P. FCS-MPC-Based Current Control of a Five-Phase Induction Motor and its Comparison with PI-PWM Control. *IEEE Trans. Ind. Electron.* **2014**, *61*, 149–163. [CrossRef]
7. Tenconi, A.; Rubino, S.; Bojoi, R. Model Predictive Control for Multiphase Motor Drives—A Technology Status Review. In Proceedings of the 2018 International Power Electronics Conference, IPEC-Niigata 2018-ECCE Asia, Niigata, Japan, 20–24 May 2018; pp. 732–739.
8. Arahal, M.R.; Barrero, F.; Ortega, M.G.; Martin, C. Harmonic analysis of direct digital control of voltage inverters. *Math. Comput. Simulat.* **2016**, *130*, 155–166. [CrossRef]
9. Aggrawal, H.; Leon, J.I.; Franquelo, L.G.; Kouro, S.; Garg, P.; Rodriguez, J. Model predictive control based selective harmonic mitigation technique for multilevel cascaded H-bridge converters. In Proceedings of the IECON 2011-37th Annual Conference of the IEEE Industrial Electronics Society, Melbourne, Australia, 7–10 November 2011; pp. 4427–4432.
10. Aguilera, R.P.; Acuña, P.; Lezana, P.; Konstantinou, G.; Wu, B.; Bernet, S.; Agelidis, V.G. Selective Harmonic Elimination Model Predictive Control for Multilevel Power Converters. *IEEE Trans. Power Electron.* **2017**, *32*, 2416–2426. [CrossRef]
11. Arahal, M.R.; Kowal, A.; Barrero, F.; del Mar Castilla, M. Cost Function Optimization for Multi-phase Induction Machines Predictive Control. *RIAI* **2018**, *16*, 1–8. [CrossRef]
12. Mamdouh, M.; Abido, M.A.; Hamouz, Z. Weighting Factor Selection Techniques for Predictive Torque Control of Induction Motor Drives: A Comparison Study. *Arab. J. Sci. Eng.* **2018**, *43*, 433–445. [CrossRef]
13. Duran, M.J.; Prieto, J.; Barrero, F.; Toral, S. Predictive Current Control of Dual Three-Phase Drives Using Restrained Search Techniques. *IEEE Trans. Ind. Electron.* **2011**, *58*, 3253–3263. [CrossRef]
14. Arahal, M.R.; Barrero, F.; Duran, M.J.; Ortega, M.G.; Martin, C. Trade-offs analysis in predictive current control of multi-phase induction machines. *Control Eng. Pract.* **2018**, *81*, 105–113. [CrossRef]

15. Gonzalez, O.; Ayala, M.; Rodas, J.; Gregor, R.; Rivas, G.; Doval-Gandoy, J. Variable-Speed Control of a Six-Phase Induction Machine using Predictive-Fixed Switching Frequency Current Control Techniques. In Proceedings of the 2018 9th IEEE International Symposium on Power Electronics for Distributed Generation Systems (PEDG), Charlotte, NC, USA, 25–28 June 2018; pp. 1–6.
16. Barrero, F.; Arahal, M.R.; Gregor, R.; Toral, S.; Duran, M.J. One-Step Modulation Predictive Current Control Method for the Asymmetrical Dual Three-Phase Induction Machine. *IEEE Trans. Ind. Electron.* **2009**, *56*, 1974–1983. [CrossRef]
17. Arahal, M.R.; Martin, C.; Barrero, F.; Gonzalez-Prieto, I.; Duran, M.J. Model-Based Control for Power Converters with Variable Sampling Time: A Case Example Using Five-Phase Induction Motor Drives. *IEEE Trans. Ind. Electron.* **2019**, *66*, 5800–5809. [CrossRef]
18. Martin, C.; Arahal, M.R.; Barrero, F.; Duran, M.J. Multiphase rotor current observers for current predictive control: A five-phase case study. *Control Eng. Pract.* **2016**, *49*, 101–111. [CrossRef]
19. Rodas, J.; Martin, C.; Arahal, M.R.; Barrero, F.; Gregor, R. Influence of Covariance-Based ALS Methods in the Performance of Predictive Controllers With Rotor Current Estimation. *IEEE Trans. Ind. Electron.* **2017**, *64*, 2602–2607. [CrossRef]
20. Levi, E. *The Industrial Electronics Handbook: Power Electronics and Motor Drives*, 2nd ed.; CRC Press: Boca Raton, FL, USA, 2011.
21. Miranda, H.; Cortes, P.; Yuz, J.I.; Rodriguez, J. Predictive Torque Control of Induction Machines Based on State-Space Models. *IEEE Trans. Ind. Electron.* **2009**, *56*, 1916–1924. [CrossRef]
22. Yepes, A.G.; Riveros, J.A.; Doval-Gandoy, J.; Barrero, F.; Lopez, O.; Bogado, B.; Jones, M.; Levi, E. Parameter Identification of Multiphase Induction Machines with Distributed Windings-Part 1: Sinusoidal Excitation Methods. *IEEE Trans. Energy Convers.* **2012**, *27*, 1056–1066. [CrossRef]
23. Riveros, J.A.; Yepes, A.G.; Barrero, F.; Doval-Gandoy, J.; Bogado, B.; Lopez, O.; Jones, M.; Levi, E. Parameter Identification of Multiphase Induction Machines with Distributed Windings-Part 2: Time-Domain Techniques. *IEEE Trans. Energy Convers.* **2012**, *27*, 1067–1077. [CrossRef]
24. IECEE. *General Guide on Harmonics and Interharmonics Measurements for Power Supply Systems and Equipment Connected Thereto*; ICE Std. 61000-4-7; IECEE: Geneva, Switzerland, 2002.

© 2019 by the authors. Licensee MDPI, Basel, Switzerland. This article is an open access article distributed under the terms and conditions of the Creative Commons Attribution (CC BY) license (http://creativecommons.org/licenses/by/4.0/).

Article

Predictive-Fixed Switching Current Control Strategy Applied to Six-Phase Induction Machine

Osvaldo Gonzalez [1,*], **Magno Ayala** [1], **Jesus Doval-Gandoy** [2], **Jorge Rodas** [1], **Raul Gregor** [1] and **Marco Rivera** [3]

1. Laboratory of Power and Control Systems (LSPyC), Facultad de Ingeniería, Universidad Nacional de Asunción, Luque 2060, Paraguay; mayala@ing.una.py (M.A.); jrodas@ing.una.py (J.R.); rgregor@ing.una.py (R.G.)
2. Applied Power Electronics Technology Research Group (APET), Universidad de Vigo, 363310 Vigo, Spain; jdoval@uvigo.es
3. Laboratory of Energy Conversion and Power Electronics, Universidad de Talca, 3340000 Curicó, Chile; marcoriv@utalca.cl
* Correspondence: ogonzalez@ing.una.py; Tel.: +59-598-370-1765

Received: 27 April 2019; Accepted: 30 May 2019; Published: 15 June 2019

Abstract: In applications such as multiphase motor drives, classical predictive control strategies are characterized by a variable switching frequency which adds high harmonic content and ripple in the stator currents. This paper proposes a model predictive current control adding a modulation stage based on a switching pattern with the aim of generating a fixed switching frequency. Hence, the proposed controller takes into account the prediction of the two adjacent active vectors and null vector in the (α-β) frame defined by space vector modulation in order to reduce the (x-y) currents according to a defined cost function at each sampling period. Both simulation and experimental tests for a six-phase induction motor drive are provided and compared to the classical predictive control to validate the feasibility of the proposed control strategy.

Keywords: multiphase induction machine; model predictive control; fixed switching frequency

1. Introduction

In recent years, multiphase induction machines (IMs) have been considered to be such a viable alternative in comparison to three-phase machines due to their fault tolerance capabilities with no extra hardware, lower torque ripple and better power splitter per phase which result very attractive to the research community for various industrial applications where a high-performance control strategy, as well as, reliability are required [1]. Presently, some applications of multiphase IMs that are being investigated include wind energy generation system [2], hybrid electric vehicles (EV) [3] and ship propulsion. In the applications mentioned above, multiphase IMs can be used under different conditions, such as healthy and post-fault operations [4,5]. From the point of view of control, the most common control strategy to regulate multiphase IMs is the field-oriented control (FOC), which is constituted by an inner current control loop, to obtain the references voltages, and an outer speed control loop for speed regulation [6]. However, several new control approaches have been carried out for the inner current control loop in multiphase IMs, some of them are: sliding mode control [7], resonant control [8] and model predictive control (MPC) [9]. Although there are other controllers such as the well-known proportional-integral (PI) controllers [10], the preferred choice is the MPC due to the fact that it shows a good transient behavior and facilitates the inclusion of nonlinearities in the system as described in [11,12], and in [13] where a comparative study between MPC and PI-PWM control has been addressed. In this context, the MPC strategy produces the reference voltage through the instantaneous discrete states of the power converter according to the minimization of a predefined cost function. However, the classic MPC strategy presents some limitations regarding to the application of only one vector in the whole sampling period. This results in current ripples as well as large voltages at low sampling frequency. Besides, the variable switching frequency develops a spread spectrum, decreasing the performance of the system in terms of useful power [14].

To overcome this subject, a predictive-fixed switching current control strategy, named (PFSCCS) from now on, applied to a two-level six-phase voltage source inverter (VSI) is presented in this paper. The strategy is based on a modulation concept employed with the MPC scheme, which has been studied for different power converters such as the mentioned two-level six-phase VSI described in [15,16] and also other topologies presented in [17,18]. In the proposed current strategy, three vectors have been considered at every sampling period, composed by two active vectors (taking only into account the largest vectors) and null vector, where their corresponding duty cycles are achieved according to the switching states and a switching pattern has also been used before being applied to VSI in order to generate a fixed switching frequency. Whereas, for the speed control loop, a PI controller has been developed by a technique shown in [19].

The main focus of this work is the implementation of the PFSCCS so as to reduce the $(x\text{-}y)$ currents compared to the classic MPC strategy using a six-phase IM supplied through a two-level six-phase VSI. In that context, both simulation and experimental validations have been included to demonstrate the capability of the proposed technique. In addition, the effectiveness of the PFSCCS is tested under steady-state and transient requirements, respectively, incorporating the mean square error (MSE) and the total harmonic distortion (THD) analysis.

The paper is organized as follows: the model of the six-phase IM and VSI are presented in Section 2. In Section 3 are described the speed controller, classic MPC and the proposed current controller based on modulated model predictive control. Section 4 shows the performance of the proposed control through simulation and experimental results in steady-state and transient conditions. Finally, Section 5 summarizes the conclusion.

2. Six-Phase IM Drive Model

The six-phase IM, supplied by a two-level six-phase VSI with a DC-Link voltage source (V_{dc}), is taken into account in this work. The simplified topology is presented in Figure 1. The six-phase IM is a dependant of time system, for this reason it is possible to represent it through a group of equations in order to define a model of the real system.

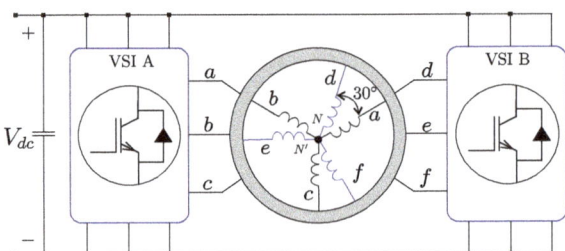

Figure 1. Six-phase IM topology supplied trough a two-level six-phase VSI.

In that sense, vector space decomposition (VSD) strategy [20] has been used to translate the actual six dimensional plane, formed through the six phases of the six-phase IM, into three two dimensional rectangular sub-spaces in the stationary reference frame, named as $(\alpha\text{-}\beta)$, $(x\text{-}y)$ and $(z_1\text{-}z_2)$ frame, by applying the amplitude invariant decoupling Clarke conversion matrix T [21]. The $(\alpha\text{-}\beta)$ frame contains the variables that provide the torque and flux regulation, unlike the $(x\text{-}y)$ frame which is linked with the energy losses. The zero elements mapped in the $(z_1\text{-}z_2)$ frame are not examined due to the adopted topology (isolated neutral points).

$$T = \frac{1}{3} \begin{pmatrix} \cos(0) & \cos(\frac{\pi}{6}) & \cos(\frac{2\pi}{3}) & \cos(\frac{5\pi}{6}) & \cos(\frac{4\pi}{3}) & \cos(\frac{3\pi}{2}) \\ \sin(0) & \sin(\frac{\pi}{6}) & \sin(\frac{2\pi}{3}) & \sin(\frac{5\pi}{6}) & \sin(\frac{4\pi}{3}) & \sin(\frac{3\pi}{2}) \\ \cos(0) & \cos(\frac{5\pi}{6}) & \cos(\frac{10\pi}{3}) & \cos(\frac{25\pi}{6}) & \cos(\frac{20\pi}{3}) & \cos(\frac{15\pi}{2}) \\ \sin(0) & \sin(\frac{5\pi}{6}) & \sin(\frac{10\pi}{3}) & \sin(\frac{25\pi}{6}) & \sin(\frac{20\pi}{3}) & \sin(\frac{15\pi}{2}) \\ 1 & 0 & 1 & 0 & 1 & 0 \\ 0 & 1 & 0 & 1 & 0 & 1 \end{pmatrix} \begin{matrix} \alpha \\ \beta \\ x \\ y \\ z_1 \\ z_2 \end{matrix} \quad (1)$$

Moreover, the model of the VSI must be included in the system. Thus, due to the discrete nature of the VSI, it is necessary to define an amount of 2^6 different switching states which represent every state of each VSI leg specified as $S_m = (S_a, ..., S_f)$, where S_m is considered as binary number, i.e., $S_m = 0$ or $S_m = 1$. Therefore, the stator phase voltages can be projected into (α-β)-(x-y) frame by considering the vector S_m and the V_{dc} voltage employing the VSD strategy. In Figure 2, the 64 control alternatives (48 active and one null vectors) are depicted in the (α-β)-(x-y) frame.

By considering the mentioned analysis, the six-phase IM can be performed by employing the state-space representation as follows:

$$x'(t) = \underbrace{\begin{pmatrix} -R_s r_2 & r_4 L_m \omega_r & 0 & 0 & r_4 R_r & r_4(L_{lr} + L_m)\omega_r \\ r_4 L_m \omega_r & -R_s r_2 & 0 & 0 & r_4(L_{lr} + L_m)\omega_r & r_4 R_r \\ 0 & 0 & -R_s r_3 & 0 & 0 & 0 \\ 0 & 0 & 0 & -R_s r_3 & 0 & 0 \\ R_s r_4 & -r_5 L_m \omega_r & 0 & 0 & -r_5 R_r & -c_5(L_{lr} + L_m) \\ -r_5 L_m \omega_r & R_s r_4 & 0 & 0 & -r_5(L_{lr} + L_m) & -r_5 R_r \end{pmatrix}}_{M_1(t)} x(t) +$$

$$\underbrace{\begin{pmatrix} r_2 & 0 & 0 & 0 \\ 0 & r_2 & 0 & 0 \\ 0 & 0 & r_3 & 0 \\ 0 & 0 & 0 & r_3 \\ -r_4 & 0 & 0 & 0 \\ 0 & -r_4 & 0 & 0 \end{pmatrix}}_{M_2(t)} u(t) + K_n v(t) \quad (2)$$

being $x(t) = (x_1, ..., x_6)^T$ the state vector constituted by stator-rotor currents of the six-phase IM, shown in Equation (3), $u(t) = (u_1, ..., u_4)^T$ is the input vector constituted by the stator voltages, presented in Equation (4). While $M_1(t)$ and $M_2(t)$ are matrices obtained by the electrical parameters of the six-phase IM. The process noise is defined as $v(t)$ and K_n represents the noise weight matrix.

$$x_1 = i_{as}, \quad x_2 = i_{\beta s}, \quad x_3 = i_{xs}, \quad x_4 = i_{ys}, \quad x_5 = i_{ar}, \quad x_6 = i_{\beta r}. \quad (3)$$

$$u_1 = u_{as}, \quad u_2 = u_{\beta s}, \quad u_3 = u_{xs}, \quad u_4 = u_{ys}. \quad (4)$$

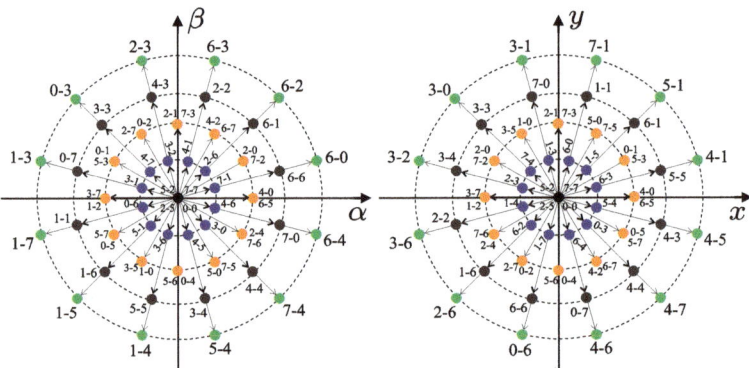

Figure 2. Mapping of the space vectors in the (α-β)-(x-y) frame for a two-level six-phase VSI.

Consequently, by taking into account the state-space representation in Equation (2) and the state vectors, it is feasible to establish the following equations:

$$\begin{aligned}
x'_1 &= -R_s r_2 x_1 + r_4 \left[L_m \omega_r x_2 + R_r x_5 + (L_{lr} + L_m) \omega_r x_6 \right] + r_2 u_1 \\
x'_2 &= -R_s r_2 x_2 + r_4 \left[-L_m \omega_r x_1 - (L_{lr} + L_m) \omega_r x_5 + R_r x_6 \right] + r_2 u_2 \\
x'_3 &= -R_s r_3 x_3 + r_3 u_3 \\
x'_4 &= -R_s r_3 x_4 + r_3 u_4 \\
x'_5 &= R_s r_4 x_1 + r_5 \left[-L_m \omega_r x_2 - R_r x_5 - (L_{lr} + L_m) \omega_r x_6 \right] - r_4 u_1 \\
x'_6 &= R_s r_4 x_2 + r_5 \left[L_m \omega_r x_1 + (L_{lr} + L_m) \omega_r x_5 - R_r x_6 \right] - r_4 u_2
\end{aligned} \quad (5)$$

where the electrical variables of the six-phase IM are represented by R_s, R_r, L_m, L_{lr} and L_{ls}, ω_r represents the rotor electrical speed and the coefficients ($r_1, ..., r_5$) are defined as:

$$r_1 = (L_{ls} + L_m)(L_{lr} + L_m) - L_m^2, \quad r_2 = \frac{L_{lr} + L_m}{r_1}, \quad r_3 = \frac{1}{L_{ls}}, \quad r_4 = \frac{L_m}{r_1}, \quad r_5 = \frac{L_{ls} + L_m}{r_1}. \quad (6)$$

Besides, in order to produce the stator phase voltages, which are dependant of the V_{dc} voltage and the vector S_m, an ideal six-phase VSI has been used [21] as it is defined in Equation (7).

$$M_{VSI} = \frac{1}{3} \begin{pmatrix} 2 & 0 & -1 & 0 & -1 & 0 \\ 0 & 2 & 0 & -1 & 0 & -1 \\ -1 & 0 & 2 & 0 & -1 & 0 \\ 0 & -1 & 0 & 2 & 0 & -1 \\ -1 & 0 & -1 & 0 & 2 & 0 \\ 0 & -1 & 0 & -1 & 0 & 2 \end{pmatrix} \begin{pmatrix} S_a \\ S_b \\ S_c \\ S_d \\ S_e \\ S_f \end{pmatrix}. \quad (7)$$

In turn, the stator phase voltages can be mapped into (α-β)-(x-y) frames defined as follows:

$$\begin{pmatrix} u_{\alpha s} \\ u_{\beta s} \\ u_{xs} \\ u_{ys} \end{pmatrix} = V_{dc} \, T \, M_{VSI} \quad (8)$$

$$\begin{pmatrix} i_{\alpha s} \\ i_{\beta s} \\ i_{xs} \\ i_{ys} \end{pmatrix} = \begin{pmatrix} 1 & 0 & 0 & 0 & 0 & 0 \\ 0 & 1 & 0 & 0 & 0 & 0 \\ 0 & 0 & 1 & 0 & 0 & 0 \\ 0 & 0 & 0 & 1 & 0 & 0 \end{pmatrix} x(t) + n(t) \quad (9)$$

where Equation (9) is considered the output vector, denoted by $y(t)$, and $n(t)$ is the measurement noise. Finally, the mechanical equations of the six-phase IM are specified as:

$$T_e = P \left(\psi_{\alpha s} i_{\beta s} - \psi_{\beta s} i_{\alpha s} \right) 3 \tag{10}$$

$$J_i \omega'_m + B_i \omega_m = (T_e - T_L) \tag{11}$$

where J_i defines the inertia coefficient, B_i is the friction coefficient, T_e represents the generated torque, T_L is the load torque, ω_m is the rotor mechanical speed, which is related to the rotor electrical speed as $\omega_r = P\omega_m$, $\psi_{\alpha s}$ and $\psi_{\beta s}$ are the stator fluxes, and P is the number of pole pairs.

3. Drive Control

A complete diagram of the PFSCCS for the six-phase IM drive is shown in Figure 3, where the outer speed control and the proposed inner current control will be detailed in the following sections.

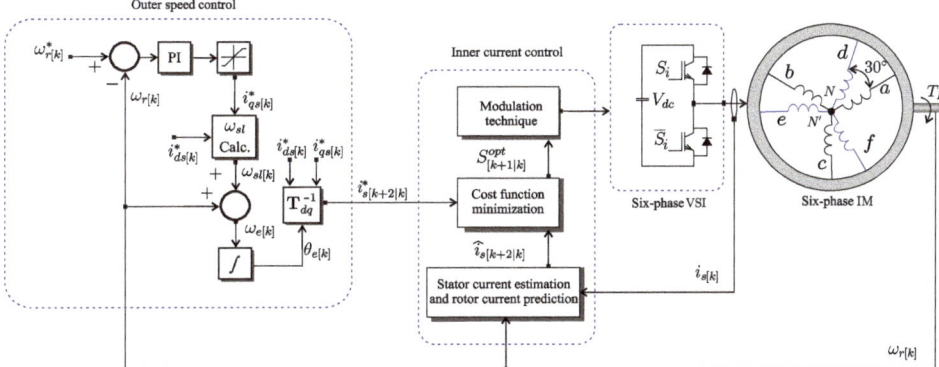

Figure 3. Complete diagram of the PFSCCS applied to six-phase IM.

3.1. Speed Control

For the external speed control loop a two degree PI controller has been incorporated, described in [19], which is based on the FOC strategy due to its easiness. Into the FOC strategy, the reference current is generated by the PI speed controller in the dynamic reference frame, known as d-q frame. Then, the reference currents are achieved by the calculation of the electric angle employed to convert the current reference, at the beginning in d-q frame, to the static reference frame (α-β), which are needed for the MPC. This method estimates the slip frequency (ω_{sl}) which is executed in the same manner as the FOC strategies, by using the reference currents in the dynamic reference frame (i^*_{ds}, i^*_{qs}) and the electrical parameters of the IM (R_r, L_r), while the rotor mechanical speed is acquired through an encoder.

3.2. Classic MPC

The MPC is related to the mathematical model of a given system, the six-phase IM in this case, commonly termed as predictive model, which consists of the prediction of the future action (at time k) of the system through measured variables, such as the rotor mechanical speed and the stator currents. Hence, for that purpose a forward Euler discretization strategy has been implemented.

$$x^p[k+1|k] = x[k] + T_s f(x[k], u[k], \omega_r[k]) \tag{12}$$

being k the actual sample and T_s the sampling period. Superscript p represents the predicted variables of the system.

According to the state-space representation (12), the stator currents and the rotor mechanical speed can be measured. Thus, the stator voltages are directly predicted through the switching states of the six-phase VSI. Nevertheless, the rotor currents are not easily measured. This issue can be faced through the estimation of the rotor currents by a reduced order estimator which determines the unmeasured fraction of the state vector. Then, in this work, the rotor currents are estimated by the proposed strategy in [22] which employs a reduced order estimator based on a Kalman Filter (KF). In that sense, uncorrelated process noises and a zero-mean Gaussian measurement have been considered. Finally, the the studied system equations are established as:

$$x^p[k+1|k] = M_1[k]x[k] + M_2[k]u[k] + K_n v[k] \tag{13}$$

$$y[k+1|k] = \begin{pmatrix} 1 & 0 & 0 & 0 & 0 & 0 \\ 0 & 1 & 0 & 0 & 0 & 0 \\ 0 & 0 & 1 & 0 & 0 & 0 \\ 0 & 0 & 0 & 1 & 0 & 0 \end{pmatrix} x[k+1] + n[k+1] \tag{14}$$

where $M_1[k]$ and $M_2[k]$ represent the discretized matrices since (5). $M_1[k]$ is related to the current value of $\omega_r[k]$ and must be included at every sampling period. A completed explanation of the aforementioned reduced order KF is presented in [22,23].

Cost Function

The optimization action is carried out at every sampling period by the MPC strategy. The action is based on the evaluation of a defined cost function, shown in (15), for every feasible stator voltages in order to obtain the control purpose. Since the cost function can be expressed in various manners, in this work, the minimization of the current tracking error has been taken into account specified by the following equation:

$$CF[k+2|k] = \sqrt{(e_{\alpha s})^2 + (e_{\beta s})^2} + \lambda_{xy}\sqrt{(e_{xs})^2 + (e_{ys})^2} \tag{15}$$

being the errors defined as follows:

$$\begin{aligned} e_{\alpha s} &= i^*_{\alpha s}[k+2] - i^p_{\alpha s}[k+2|k], \\ e_{\beta s} &= i^*_{\beta s}[k+2] - i^p_{\beta s}[k+2|k], \\ e_{xs} &= i^*_{xs}[k+2] - i^p_{xs}[k+2|k], \\ e_{ys} &= i^*_{ys}[k+2] - i^p_{ys}[k+2|k]. \end{aligned} \tag{16}$$

considering $i^*_s[k+2]$ as the reference vector for the stator currents and $i^p_s[k+2]$ as the predicted values based on the second-step forward state. At the same time, a tuning parameter is included in the cost function, described as λ_{xy}, in order to provide an extra weight on (α-β) or (x-y) frames [22,23].

3.3. Proposed Current Controller (Pfsccs)

According to the space vector modulation (SVM) strategy, it is feasible to find the available vectors for the six-phase VSI in the (α-β) frame, this produces 64 sectors (48 different ones), which are conformed by two active vectors and a null vector as depicted in Figure 4. The proposed strategy realizes the prediction of the vectors (null vector and two active vectors) that compose every sectors and calculates the cost function independently (G_0, G_1 and G_2) for each prediction at every sampling period. However, the proposed strategy only select the twelve largest vectors including the null vector in order to represent the optimal vector. This current control approach has been adopted in order to reduce the (x-y) currents [24,25].

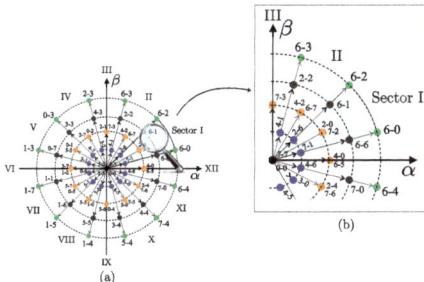

Figure 4. Considered sectors for the six-phase VSI in the (α-β) frame: (**a**) Available vectors; (**b**) A selected specific sector shown as zoom.

The prediction is obtained by Equation (13), but differs in the calculation of the input vector ($u[k]$) [21]. The duty cycles (d_c), considering the null vector and the two active vectors (d_{c-0}, d_{c-1} and d_{c-2}), respectively, are achieved through the following equations:

$$d_{c-0} = \frac{\mu}{G_0}, \quad d_{c-1} = \frac{\mu}{G_1}, \quad d_{c-2} = \frac{\mu}{G_2}, \tag{17}$$

$$d_{c-0} + d_{c-1} + d_{c-2} = 1, \tag{18}$$

Hence, it is possible to acquire the relation for μ and the duty cycles for the specified vectors as:

$$d_{c-0} = \frac{G_1 G_2}{G_0 G_1 + G_1 G_2 + G_0 G_2}, \tag{19}$$

$$d_{c-1} = \frac{G_0 G_2}{G_0 G_1 + G_1 G_2 + G_0 G_2}, \tag{20}$$

$$d_{c-2} = \frac{G_0 G_1}{G_0 G_1 + G_1 G_2 + G_0 J_2}. \tag{21}$$

Taking account these relations, the cost function is redefined, as shown in Equation (22), and calculated at each T_s.

$$CF_n[k+2|k] = d_{c-1} G_1 + d_{c-2} G_2. \tag{22}$$

In this way, the two vectors that reduce $CF_n[k+2|k]$ are chosen and then applied to the VSI at the following sampling period. Once the optimal vectors are obtained, the two active vectors (v_1-v_2) and null vector (v_0), their respective duty cycles to be applied and a switching pattern scheme, described in [21], are taken with the aim of producing a fixed switching frequency.

4. Simulation and Experimental Results

First, simulations have been performed in a MATLAB/Simulink R2014a environment so as to verify the feasibility of the PFSCCS using a six-phase IM shown in Figure 1. Numerical integration using first order Euler's algorithm has been applied to calculate the progress of the studied system. The simulation parameters of the six-phase IM are listed in Table 1.

Table 1. Characteristics of the six-phase IM.

R_r	6.9 Ω	L_s	654.4 mH
L_{lr}	12.8 mH	L_r	626.8 mH
L_{ls}	5.3 mH	P_w	2 kW
R_s	6.7 Ω	J_i	0.07 kg.m²
L_m	614 mH	B_i	0.0004 kg.m²/s
P	1	ω_{r-nom}	3000 rpm

The effectiveness of the presented control technique for the six-phase IM has been evaluated under a load condition ($T_L = 2$ Nm), the sampling frequency is 8 kHz, V_{dc} is 400 V and the d-axis current reference (i_{ds}^*) has been set in 1 A, while for the gains of the two degree PI controller with a saturation, can be found in [19]. Moreover, for the proposed control, $\lambda_{xy} = 0.1$, defined in (15), has been used in order to give more emphasis to the (α-β) stator current tracking.

The performance of the proposed technique is compared in transient and steady-state conditions. Both proofs, simulation and experimental results, are analyzed in terms of mean squared error (MSE) and total harmonic distortion (THD) obtained between the reference and the measured stator currents in the (α-β) and (x-y) sub-spaces for MSE test and the THD is obtained in the (α-β) sub-space. The MSE is computed as follows:

$$\text{MSE}(i_{\sigma s}) = \sqrt{\frac{1}{N} \sum_{j=1}^{N} (i_{\sigma s} - i_{\sigma s}^*)^2} \tag{23}$$

where the stator current reference is represented through the superscript $*$, the measured stator current is defined by $i_{\sigma s}$ taking into account that σ includes the (α-β)-(x-y) frame and N is the number of studied samples. While, the THD is obtained as follows:

$$\text{THD}(i_s) = \sqrt{\frac{1}{i_{s1}^2} \sum_{k=2}^{N} (i_{sk})^2} \tag{24}$$

where i_{s1} corresponds to the fundamental stator current whereas i_{sk} is the harmonic stator current (multiple of the fundamental stator current).

In Figure 5 the performance of the stator currents in the (α-β)-(x-y) frame can be seen in steady-state condition. According to the simulations results, shown in Table 2, the proposed technique has a good behavior considering the MSE and THD analysis of the stator currents at different rotor mechanical speeds. In addition, it can be noticed that at lower speeds, the stator currents ripple in the (α-β) frame is slightly smaller than at higher mechanical rotor speeds, in the same way that occurs for the (x-y) currents.

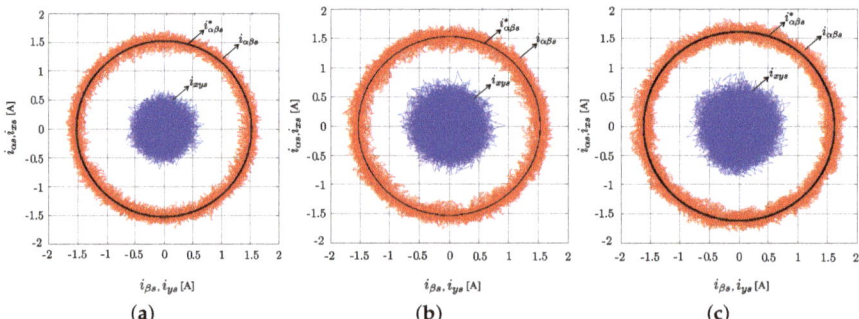

Figure 5. Simulation performance in steady-state condition of stator currents in (α-β) and (x-y) sub-spaces for a sampling frequency of 8 kHz at different speeds (ω_m): (**a**) 500 rpm; (**b**) 1000 rpm; (**c**) 1500 rpm.

Table 2. Simulation performance test of stator currents (α-β), (x-y), MSE (A), THD (%) at different rotor speeds (rpm).

			f_s = 8 kHz			
ω_m^*	MSE_α	MSE_β	MSE_x	MSE_y	THD_α	THD_β
500	0.065	0.064	0.174	0.172	5.73	5.46
1000	0.076	0.075	0.211	0.203	5.43	5.34
1500	0.110	0.110	0.219	0.216	6.46	6.38

For the experimental proofs the PFSCCS, previously described, is examined in the test rig shown in Figure 6 in order to prove its effectiveness, employing a six-phase IM supplied through two tradictional three-phase VSI, being analogous to a six-phase VSI and the V_{dc} voltage is obtained by means of a DC power source. A dSPACE MABXII DS1401 real-time rapid prototyping bench including Simulink version 8.2 has been used to manage the two-level six VSI. Once the results are acquired, these have been analyzed through MATLAB/Simulink R2014a code. Employing stand-still with VSI proofs and AC time domain strategies, the electrical parameters have been acquired [26,27]. Table 1 summarizes these results. Current sensors LA 55-P s (frequency bandwidth since DC up to 200 kHz) have been used for the experimental measurements. The current measurements have been then turned to digital format by means of a 16-bit A/D converter. The six-phase IM angle has been measured with a 1024-pulses per revolution (ppr) incremental encoder in order to estimate the rotor speed and also a 5 HP eddy current brake has been used to insert a fixed mechanical load on the system.

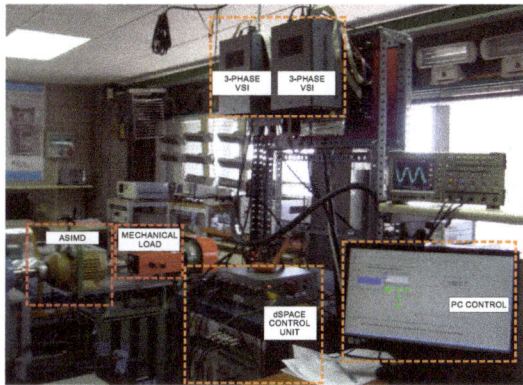

Figure 6. Experimental test rig.

Taking this into account, the experimental results have been analyzed with the same tests that simulations results as figures of merit. The stator currents reference in the (x-y) frame have been established to zero, i.e., $i_{xs}^* = i_{ys}^* = 0$ A so as to decrease the losses in the copper. The amounts for the process noise ($\hat{Q}_w = 0.0022$) and the measurement noise ($\hat{R}_v = 0.0022$) is estimated by means of the strategy proposed in [23]. The dynamic behavior of the proposed technique has been evaluated with two different values of λ_{xy}, defined in (15), giving more weight to (α-β) stator currents tracking. In the developed tests, the sampling frequencies have been fixed in 8 kHz for PFSCCS and 8 kHz and 16 kHz for classic MPC, respectively, due to the fact that the PFSCCS uses two active vectors and null vector twice in a sampling period and this procedure doubles the switching frequency compared to the sampling frequency. In that sense, tests have been included in order to expose a fair comparison between the classic MPC and PFSCCS at the mentioned sampling frequencies and also to show the performance of both techniques. For the rotor mechanical speeds, two operation points have been considered, 500 rpm and 1000 rpm, respectively, in steady-state condition. Furthermore, for a transient response, a reversal rotor mechanical speed test from 500 rpm to -500 rpm has been considered for PFSCCS and from 1500 rpm to 200 rpm for classic MPC and PFSCCS. The obtained results between classic MPC and PFSCCS are reported in Table 3, where the proposed current control technique has demonstrated a good tracking of the current references considering the MSE and THD in the (α-β)-(x-y) frame.

Table 3. Experimental performance test of stator currents (α-β), (x-y), MSE (A), THD (%) between classic MPC and the PFSCCS at different rotor speeds (rpm).

ω_m^*	MSE_α	MSE_β	MSE_x	MSE_y	THD_α	THD_β
\multicolumn{7}{c}{f_s = 8 kHz for Classic MPC}						
500	0.140	0.130	0.821	0.822	8.30	8.40
1000	0.147	0.138	0.953	0.934	7.40	7.30
\multicolumn{7}{c}{f_s = 16 kHz for Classic MPC}						
500	0.073	0.072	0.491	0.483	8.40	8.30
1000	0.084	0.082	0.538	0.534	7.50	7.40
\multicolumn{7}{c}{f_s = 8 kHz for PFSCCS}						
500	0.042	0.045	0.135	0.130	4.89	5.08
1000	0.069	0.068	0.197	0.204	4.69	4.78

Figure 7 presents the trajectories of the stator currents in the (α-β)-(x-y) frame of the PFSCCS applied to the six-phase IM. In this test two different values of the tuning parameter (λ_{xy}) have been considered, in Figure 7a, λ_{xy} = 0.05 has been considered and λ_{xy} = 0.1 in Figure 7b. The rotor mechanical speed has been set to 500 rpm at 8 kHz. The figure shows that (x-y) currents decrease when λ_{xy} increases, which imply that the selection of this parameter has a strong influence on the behavior of the system. Further, the (α-β) current tracking has a slightly better performance considering λ_{xy} = 0.1.

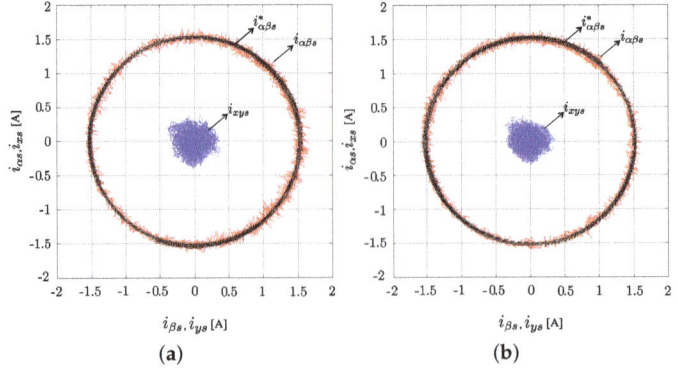

Figure 7. Experimental results in the (α-β)-(x-y) frame for stator currents at 8 kHz of sampling frequency and 500 rpm rotor speed considering: (**a**) λ_{xy} = 0.05; (**b**) λ_{xy} = 0.1.

In addition, Figure 8a shows the harmonic content of the measured stator current ($i_{\alpha s}$) through THD analysis and also, in Figure 8b has been included the switching voltage in the six-phase VSI showing the pattern of the proposed modulation strategy.

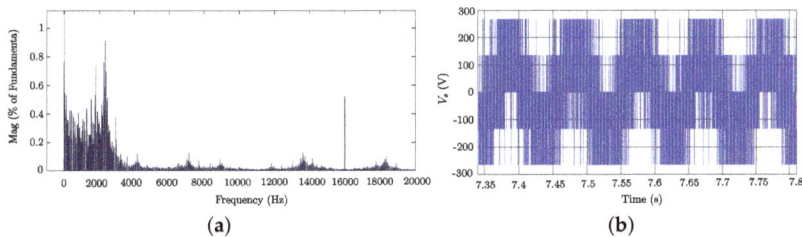

Figure 8. Experimental performance for PFSCCS at 8 kHz of sampling frequency and 500 rpm: (**a**) Spectrum of the measured stator current; (**b**) Switching pattern in the VSI.

On the other hand, Figure 9 exposes the transient response of the proposed control for a step response in q axis. The transient response has been included through a reversal test from rotor mechanical speed (500 rpm to −500 rpm) at 8 kHz. Both cases report fast responses considering the overshoot and settling time, which were of 6.14% and 6 ms, respectively, for Figure 9a and 4.85% and 6.12 ms, respectively for Figure 9b. The criterion of the 5% has been selected. Finally, a experimental transient response from a step change of 1500 rpm to 200 rpm between classic MPC and PFSCCS has been depicted in Figure 10 in order to show the performance of the proposed strategy, which it has demonstrated that it can be used in industrial applications (e.g., regenerating braking).

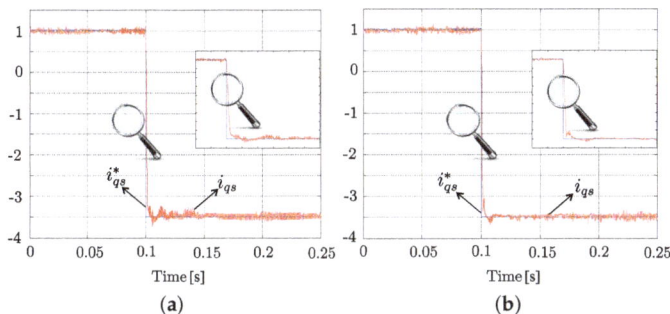

Figure 9. Experimental transient test in q-axis of stator currents from a speed change of 500 rpm to −500 rpm at 8 kHz of sampling frequency considering: (**a**) $\lambda_{xy} = 0.05$; (**b**) $\lambda_{xy} = 0.1$.

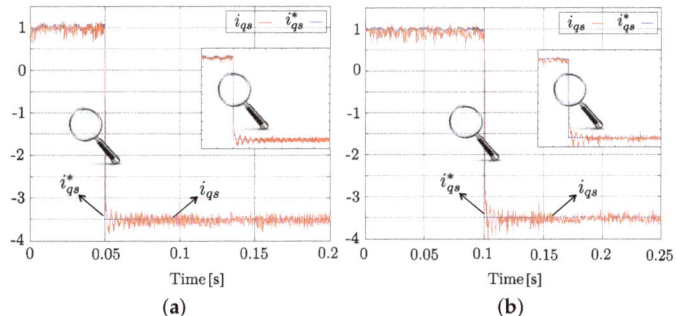

Figure 10. Experimental transient test in q-axis of stator currents from a speed change of 1500 rpm to 200 rpm at 16 kHz and 8 kHz of sampling frequency, respectively: (**a**) Classic MPC; (**b**) PFSCCS.

5. Conclusions

In this paper, a predictive current control technique with a fixed switching frequency applied to a six-phase IM has been proposed. This technique has been developed to reduce the stator currents in the $(x\text{-}y)$ frame using the largest vectors of the $(\alpha\text{-}\beta)$ frame with a stage of modulation based on a determined switching pattern in order to produce a fixed switching frequency. The simulation and experimental results have shown the performance of the proposed technique, where the system has been tested under different conditions (steady and transient conditions) including different rotor mechanical speeds, sampling frequency and tuning parameters for $(x\text{-}y)$ stator currents, respectively. In terms of $(\alpha\text{-}\beta)$ currents tracking, the presented technique has a better behavior at lower speed and a remarkable reduction of $(x\text{-}y)$ stator currents compared to classic MPC. The obtained results have also demonstrated a good transient current behavior in terms of overshoot and settling time. In summary, the proposed current control technique is a good alternative both in low and high speeds for industrial applications.

Author Contributions: Conceptualization, O.G., M.A. and J.R.; methodology, O.G., M.A.; software, M.A. and O.G.; validation, O.G., M.A. and J.D.-G.; formal analysis, O.G., M.A., J.D.-G., J.R., R.G. and M.R.; investigation, O.G. and M.A.; resources, J.D.-G.; data curation, J.D.-G., O.G. and M.A.; writing, original draft preparation, O.G., M.A. and J.R.; writing, review and editing, O.G., M.A., J.R., J.D.-G., R.G. and M.R.; visualization, J.D.-G., M.A., J.R., R.G. and M.R.; project administration, O.G., M.A., J.R., and J.D.-G.; funding acquisition, R.G. and J.R.

Funding: This research has been funded through the Consejo Nacional de Ciencia y Tecnología (CONACYT)-Paraguay, Grant Number POSG16-05 and FONDECYT Regular 1160690 Research Project.

Acknowledgments: The authors would like to thank members of the LSPyC, APET and LECPE for their significant comments on this work.

Conflicts of Interest: The authors declare no conflict of interest.

Abbreviations

The following abbreviations have been employed in this work:

FOC	Field Oriented Control
IM	Induction Machine
MPC	Model Predictive Control
PFSCCS	Predictive-Fixed Switching Current Control Strategy
MSE	Mean Squared Error
PI	Proportional-Integral
PWM	Pulse-Width Modulation
SVM	Space Vector Modulation
THD	Total Harmonic Distortion
VSI	Voltage Source Inverter
VSD	Vector Space Decomposition

References

1. Duran, M.J.; Levi, E.; Barrero, F. Multiphase Electric Drives: Introduction. In *Wiley Encyclopedia of Electrical and Electronics Engineering*. Available online: https://onlinelibrary.wiley.com/doi/abs/10.1002/047134608X.W8364 (accessed on 26 April 2019).
2. Precup, R.E.; Kamal, T.; Hassan, S.Z. *Advanced Control and Optimization Paradigms for Wind Energy Systems*; Springer: Berlin/Heidelberg, Germany, 2019.
3. Subotic, I.; Bodo, N.; Levi, E. Integration of six-phase EV drivetrains into battery charging process with direct grid connection. *IEEE Trans. Energy Conv.* **2017**, *32*, 1012–1022. [CrossRef]
4. Ayala, M.; Gonzalez, O.; Rodas, J.; Gregor, R.; Doval-Gandoy, J. A speed-sensorless predictive current control of multiphase induction machines using a Kalman filter for rotor current estimator. In Proceedings of the 2016 International Conference on Electrical Systems for Aircraft, Railway, Ship Propulsion and Road Vehicles & International Transportation Electrification Conference (ESARS-ITEC), Toulouse, France, 2–4 November 2016; pp. 1–6.

5. Munim, W.N.W.A.; Duran, M.J.; Che, H.S.; Bermúdez, M.; González-Prieto, I.; Rahim, N.A. A unified analysis of the fault tolerance capability in six-phase induction motor drives. *IEEE Trans. Power Electron.* **2017**, *32*, 7824–7836. [CrossRef]
6. Jones, M.; Vukosavic, S.N.; Dujic, D.; Levi, E. A synchronous current control scheme for multiphase induction motor drives. *IEEE Trans. Energy Conv.* **2009**, *24*, 860–868. [CrossRef]
7. Kali, Y.; Ayala, M.; Rodas, J.; Saad, M.; Doval-Gandoy, J.; Gregor, R.; Benjelloun, K. Current Control of a Six-Phase Induction Machine Drive based on Discrete-Time Sliding Mode with Time Delay Estimation. *Energies* **2019**, *12*, 170. [CrossRef]
8. Che, H.S.; Duran, M.J.; Levi, E.; Jones, M.; Hew, W.P.; Rahim, N.A. Postfault operation of an asymmetrical six-phase induction machine with single and two isolated neutral points. *IEEE Trans. Power Electron.* **2014**, *29*, 5406–5416. [CrossRef]
9. Mirzaeva, G.; Goodwin, G.C.; McGrath, B.P.; Teixeira, C.; Rivera, M. A generalized MPC framework for the design and comparison of VSI current controllers. *IEEE Trans. Ind. Electron.* **2016**, *63*, 5816–5826. [CrossRef]
10. Vazquez, S.; Rodriguez, J.; Rivera, M.; Franquelo, L.G.; Norambuena, M. Model Predictive Control for Power Converters and Drives: Advances and Trends. *IEEE Trans. Ind. Electron.* **2016**, *64*, 935–947. [CrossRef]
11. Barrero, F.; Arahal, M.R.; Gregor, R.; Toral, S.; Durán, M.J. A proof of concept study of predictive current control for VSI-driven asymmetrical dual three-phase AC machines. *IEEE Trans. Ind. Electron.* **2009**, *56*, 1937–1954. [CrossRef]
12. Barrero, F.; Prieto, J.; Levi, E.; Gregor, R.; Toral, S.; Durán, M.J.; Jones, M. An enhanced predictive current control method for asymmetrical six-phase motor drives. *IEEE Trans. Ind. Electron.* **2011**, *58*, 3242–3252. [CrossRef]
13. Lim, C.S.; Levi, E.; Jones, M.; Rahim, N.A.; Hew, W.P. FCS-MPC-based current control of a five-phase induction motor and its comparison with PI-PWM control. *IEEE Trans. Ind. Electron.* **2013**, *61*, 149–163. [CrossRef]
14. Vijayagopal, M.; Zanchetta, P.; Empringham, L.; De Lillo, L.; Tarisciotti, L.; Wheeler, P. Modulated model predictive current control for direct matrix converter with fixed switching frequency. In Proceedings of the 2015 17th European Conference on Power Electronics and Applications (EPE'15 ECCE-Europe), Geneva, Switzerland, 8–10 September 2015; pp. 1–10.
15. Gregor, R.; Rodas, J.; Munoz, J.; Ayala, M.; Gonzalez, O.; Gregor, D. Predictive-Fixed Switching Frequency Technique for 5-Phase Induction Motor Drives. In Proceedings of the 2016 International Symposium on Power Electronics, Electrical Drives, Automation and Motion (SPEEDAM), Anacapri, Italy, 22–24 June 2016.
16. Ayala, M.; Gonzalez, O.; Rodas, J.; Gregor, R.; Rivera, M. Predictive control at fixed switching frequency for a dual three-phase induction machine with Kalman filter-based rotor estimator. In Proceedings of the 2016 IEEE International Conference on Automatica (ICA-ACCA), Curico, Chile, 19–21 October 2016; pp. 1–6.
17. Rivera, M.; Toledo, S.; Baier, C.; Tarisciotti, L.; Wheeler, P.; Verne, S. Indirect predictive control techniques for a matrix converter operating at fixed switching frequency. In Proceedings of the 2017 IEEE International Symposium on Predictive Control of Electrical Drives and Power Electronics (PRECEDE), Pucón, Chile, 18–20 October 2017; pp. 13–18.
18. Comparatore, L.; Gregor, R.; Rodas, J.; Pacher, J.; Renault, A.; Rivera, M. Model based predictive current control for a three-phase cascade H-bridge multilevel STATCOM operating at fixed switching frequency. In Proceedings of the 2017 IEEE 8th International Symposium on Power Electronics for Distributed Generation Systems (PEDG), Florianopolis, Brazil, 17–20 April 2017; pp. 1–6.
19. Harnefors, L.; Saarakkala, S.; Hinkkanen, M. Speed Control of Electrical Drives Using Classical Control Methods. *IEEE Trans. Ind. Appl.* **2013**, *49*, 889–898. [CrossRef]
20. Zhao, Y.; Lipo, T. Space vector PWM control of dual three-phase induction machine using vector space decomposition. *IEEE Trans. Ind. Electron.* **1995**, *31*, 1100–1109.
21. Ayala, M.; Rodas, J.; Gregor, R.; Doval-Gandoy, J.; Gonzalez, O.; Saad, M.; Rivera, M. Comparative Study of Predictive Control Strategies at Fixed Switching Frequency for an Asymmetrical Six-Phase Induction Motor Drive. In Proceedings of the 2017 IEEE International Electric Machines and Drives Conference (IEMDC), Miami, FL, USA, 21–24 May 2017; pp. 1–8.
22. Rodas, J.; Barrero, F.; Arahal, M.R.; Martin, C.; Gregor, R. On-Line Estimation of Rotor Variables in Predictive Current Controllers: A Case Study Using Five-Phase Induction Machines. *IEEE Trans. Ind. Electron.* **2016**, *63*, 5348–5356. [CrossRef]

23. Rodas, J.; Martin, C.; Arahal, M.R.; Barrero, F.; Gregor, R. Influence of Covariance-Based ALS Methods in the Performance of Predictive Controllers with Rotor Current Estimation. *IEEE Trans. Ind. Electron.* **2017**, *64*, 2602–2607. [CrossRef]
24. Pandit, J.K.; Aware, M.V.; Nemade, R.V.; Levi, E. Direct torque control scheme for a six-phase induction motor with reduced torque ripple. *IEEE Trans. Power Electron.* **2017**, *32*, 7118–7129. [CrossRef]
25. Gonzalez, O.; Ayala, M.; Rodas, J.; Gregor, R.; Rivas, G.; Doval-Gandoy, J. Variable-Speed Control of a Six-Phase Induction Machine using Predictive-Fixed Switching Frequency Current Control Techniques. In Proceedings of the 2018 9th IEEE International Symposium on Power Electronics for Distributed Generation Systems (PEDG), Charlotte, NC, USA, 25–28 June 2018; pp. 1–6.
26. Yepes, A.G.; Riveros, J.A.; Doval-Gandoy, J.; Barrero, F.; López, O.; Bogado, B.; Jones, M.; Levi, E. Parameter identification of multiphase induction machines with distributed windings Part 1: Sinusoidal excitation methods. *IEEE Trans. Energy Conv.* **2012**, *27*, 1056–1066. [CrossRef]
27. Riveros, J.A.; Yepes, A.G.; Barrero, F.; Doval-Gandoy, J.; Bogado, B.; Lopez, O.; Jones, M.; Levi, E. Parameter identification of multiphase induction machines with distributed windings Part 2: Time-domain techniques. *IEEE Trans. Energy Conv.* **2012**, *27*, 1067–1077. [CrossRef]

© 2019 by the authors. Licensee MDPI, Basel, Switzerland. This article is an open access article distributed under the terms and conditions of the Creative Commons Attribution (CC BY) license (http://creativecommons.org/licenses/by/4.0/).

Article

Constraint Satisfaction in Current Control of a Five-Phase Drive with Locally Tuned Predictive Controllers

Agnieszka Kowal G. [1], Manuel R. Arahal [1,*], Cristina Martin [2] and Federico Barrero [2]

[1] Systems Engineering and Automation Department, University of Seville, 41092 Seville, Spain
[2] Electronic Engineering Department, University of Seville, 41092 Seville, Spain
* Correspondence: arahal@us.es; Tel.: +34-954-48-73-43

Received: 9 May 2019; Accepted: 10 July 2019; Published: 16 July 2019

Abstract: The problem of control of stator currents in multi-phase induction machines has recently been tackled by direct digital model predictive control. Although these predictive controllers can directly incorporate constraints, most reported applications for stator current control of drives do no use this possibility, being the usual practice tuning the controller to achieve the particular compromise solution. The proposal of this paper is to change the form of the tuning problem of predictive controllers so that constraints are explicitly taken into account. This is done by considering multiple controllers that are locally optimal. To illustrate the method, a five-phase drive is considered and the problem of minimizing $x-y$ losses while simultaneously maintaining the switching frequency and current tracking error below some limits is tackled. The experiments showed that the constraint feasibility problem has, in general, no solution for standard predictive control, whereas the proposed scheme provides good tracking performance without violating constraints in switching frequency and at the same time reducing parasitic currents of $x-y$ subspaces.

Keywords: constraints satisfaction; cost functions; local controllers; predictive current control; multi-phase drives

1. Introduction

This papers deals with stator current control of Induction Machines (IM) with more than three phases. It is well known that Model Based Predictive Control (MBPC) can be applied for this task in a configuration where the controller directly commands the inverter without modulation techniques [1]. This control scheme is a particular case of the more general Finite State Model Predictive Control and is often referred to as Predictive Current Control (PCC) [2–4]. Recently, PCC has received increasing attention as an interesting choice for multi-phase and/or multi-level systems.

One of the key aspects of MBPC is the possibility of handling constraints directly [5] and, thus, the PCC could benefit from the constraint-handling capability of MBPC, however in most reported cases this possibility is not used. Instead, the usual practice is tuning the controller to achieve the particular compromise solution [6]. Thus, the selection of the controller parameters is made based on the expected behavior of the IM considering some operating regions. It must be recalled that in PCC the instantaneous minimization of the cost function imprints in the IM certain current waveforms that in turn produce a a certain global behavior. It is often found that such behavior contains conflicting criteria, thus PCC design should meet the underlying trade-offs [7]. According to this, PCC tuning should translate objectives such as commutations, tracking quality, etc. to control parameters, which is not an easy task. Regarding this, several methods have been proposed to tune the MBPC for drives in a more or less automatic fashion (see [8,9] for a review of methods), but the constraint satisfaction is not considered.

The proposal of this paper is to change the form of the tuning problem of MBPC so that constraints are explicitly taken into account. It is shown that this can be done by considering multiple controllers that are locally optimal in a way similar to the proposal in [10], where, by solving the optimization problem differently for each operating region, a better global behavior can be attained.

To illustrate the method, in this paper a five-phase drive is considered. The higher number of phases (compared to the standard three-phase case) provides further room for optimization, more tuning possibilities and complex trade-offs between the different figures of merit. For this application, the problem of minimizing $x - y$ losses while simultaneously maintaining the switching frequency and current tracking error below some limits is considered. For other applications, other figures of merit could be used, for instance harmonic distortion in Uninterrupted Power Systems [11], current ripple in permanent magnet motors [12], speed in wind turbines [13] and others.

The chosen example problem is relevant as the five-phase machine is of interest [14] and the proposed minimization would reduce losses without compromising dynamic performance and ensuring a switching frequency adequate for the available hardware. Please notice that the strategy is applicable to other types of systems, being the five-phase IM a particularly interesting and demanding case study that requires dealing with the extra number of phases. Application to other multi-phase systems such as the six-phase IM is straightforward thanks to the vector space decomposition technique [15].

In the next section, the basic aspects of PCC are reviewed, introducing the figures of merit that are considered in the proposal for MBPC tuning. The concept of constraint handling via local controllers is presented in Section 3 including the partition of operating space and the local tuning method. The resulting controller is assessed in Section 4, paying special attention to the constraint feasibility problem for the whole operating space. From these results, some conclusions are derived in Section 5.

2. PCC for Five-Phase IM

A brief introduction to PCC is now given to ease the presentation of the proposed variation. Although the description is given for the specific case of a five-phase IM, only minor adjustment are needed for a different number of phases thanks to the state-space representation.

According to the block diagram of PCC shown in Figure 1, the IM is driven by a Voltage Source Inverter (VSI) that provides a certain set of phase voltages v that are derived from the control signal u and the VSI topology. This produces stator phase currents i that follow a vector of sinusoidal reference trajectories i^*. The PCC uses a model of the IM and VSI to predict the evolution of the stator currents associated to each possible VSI state. The controller optimizes, by exhaustive search, the control move for discrete time $k + 1$, which is held for a sampling period. The procedure is then repeated according to the receding horizon rule typical of predictive controllers.

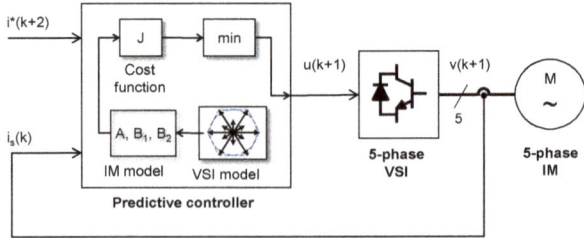

Figure 1. Diagram of predictive stator current control of a five-phase IM fed by a VSI.

In the case of multi-phase IM with n phases, the vector space decomposition technique provides the decomposition of the n-phase space into one $\alpha - \beta$ plane, which is responsible for energy conversion and some other planes: $(x - y)^1$ to $(x - y)^{(n-4)/2}$ that are related to copper losses. The voltages produced by the VSI are mapped to these subspaces to form a row vector $v_{\alpha\beta xys} = (v_{\alpha s}, v_{\beta s}, v_{xs}, v_{ys})$ by means of

$$v_{\alpha\beta xys} = V_{dc} u T M \tag{1}$$

where V_{dc} is the DC link voltage, T is a connectivity matrix that takes into account how the VSI gating signals are distributed and M is a coordinate transformation matrix accounting for the spatial distribution of machine windings. In the case of a three-phase IM, there is no $x - y$ subspace. In the case of the five-phase IM, only one auxiliary plane exists. For other multi-phase systems, the extra number of subspaces are easily treated using the state-space representation. From this decomposition, and using time-discretization, the following predictive model is obtained

$$\hat{i}(k+2|k) = Ai(k) + B_1 u(k) + B_2 u(k+1) + G(k) \tag{2}$$

where matrices A, B_1 and B_2 are related to the IM electrical parameters and to the VSI connections. In addition, matrix A is not constant but dependent on the electrical frequency $A(\omega_r)$ [16]. A state space vector can be considered including stator currents in the principal and secondary planes such as $i = (i_\alpha, i_\beta, i_x, i_y)^\top$. This vector can be obtained from sensors transforming the phase stator currents i_s into $\alpha - \beta$ and $x - y$ subspaces by means of the inverse transformation to that given in Equation (1). Vector G accounts for the dynamics due to rotor currents that are usually not measured. It constitutes a term that must be estimated at each k [17,18].

In Equation (2), the control signal u is a vector of gating signals of the VSI $u = (K_1, K_2, \ldots, K_5)^\top$, where $K_i \in \{0, 1\}$ for $i = 1, \ldots, 5$. Each of the five phases can be connected through the VSI to the positive ($K_i = 1$) or negative ($K_i = 0$) borne of the DC-link. In this way, the VSI can produce $\tau = 2^5$ different phase voltages. Due to redundancy, only 31 different voltages are actually produced [19].

In most applications of PCC, for discrete time k, a value of the control signal is selected for the next sampling period $u(k+1)$ by minimizing some objective function J. This objective function can be the sum of a number of quadratic terms penalizing control error, control effort, etc. [5]. The simplest objective function includes the square of the predicted control error $\hat{e} = (i^* - \hat{i})$. In the case of drives, sometimes a penalization of VSI commutations is included. This is achieved by computing the number of switch changes SC produced at the VSI when the previous state $u(k)$ is changed to any other $u(k+1)$ as $SC(k) = \sum_{i=1}^{5} |u_i(k+1) - u_i(k)|$, and the switching frequency f_{sw} is $f_{sw} = \sum_{k=k_1}^{k_2} SC(k)/(T_s(k_2 - k_1 + 1))$, where T_s is the sampling time. Please notice that the switching frequency in PCC is not constant. This drawback is a consequence of the direct digital control approach. Although this can be alleviated by using some techniques, the benefits from the elimination of the modulation stage are deemed more important in most cases. Nevertheless, the average switching frequency should not exceed some limits imposed by the VSI hardware. In addition, to limit commutation losses high values of f_{sw} should be avoided. Finally, the $x - y$ currents do not produce torque and are related to losses, thus it is customary to set their reference to zero. Taking these considerations into account, the cost function takes the form

$$J = \|\hat{e}_{\alpha\beta}\|^2 + \lambda_{xy} \|\hat{e}_{xy}\|^2 + \lambda_{sc} SC \tag{3}$$

where $\|.\|$ is the vector modulus operation. It can be seen that two parameters (λ_{xy} and λ_{sc}) are needed to take into account the different scales of the variables included in the cost function. In addition, these parameters are typically used to set the relative importance of the three objectives. In a traditional PCC setup, these factors are computed off-line as a compromise between conflicting criteria and considering (in the best of cases) the whole range of operation of the system, as in [6]. The computed values are then used on-line. This way of proceeding seems subject to potential improvement as it is realized that, for different regions of operation, the terms present in Equation (3) take different relative values. For instance, for low speed and load, the IM shows larger harmonic distortion and lower switching frequency, whereas, for medium speed and load, the situation is the opposite [7].

3. Constraint Handling with Local Controllers

The problem associated with constraints in PCC is that the minimization of Equation (3) does not guarantee a certain quality of tracking or a certain commutation rate. This is because the minimization of Equation (3) takes place at each sampling time without regard the choices made in previous periods. A possible path to overcome this would be determining a set of controller parameters (λ_{xy} and λ_{sc}) to attain the desired behavior. The following optimization problem can then be used

$$\min_{\theta} \quad E_{xy}$$
$$\text{s.t.} \quad E_{\alpha\beta} < U_{\alpha\beta} \quad (4)$$
$$\max f_{sw} < U_{sw}$$

where $\theta = (\lambda_{xy}, \lambda_{SC})$ is the generic element of the search space, $E_{\alpha\beta}$ is the root mean squared error obtained for tracking of $\alpha - \beta$ stator currents, E_{xy} is the root mean squared error obtained for regulation of $x - y$ stator currents, and $\max f_{sw}$ is the highest recorded value for the switching frequency f_{sw}.

With this design procedure a minimization of $x - y$ related losses is sought ensuring at the same time that the tracking error is below some limit $U_{\alpha\beta}$ and that the VSI would not exceed a limit U_{sw} set on the switching frequency f_{sw}. Unfortunately, this problem, in general, not solvable for all operating regimes because no feasible solutions exist [7]. It is shown that using locally tuned controllers it is possible to attain a solution for each operating regime.

3.1. Local Controllers

The concept of Local Controllers is similar to other well known methods such as the Self Tuning Regulator [20] and Gain Scheduling [21]. The original ideas of dividing a non-linear design problem into linear sub-problems have spurred many variations including the strategy known as Local Controller Networks [22] where a set of controllers is considered instead of just a fixed one. At any given moment, only one controller from the set is allowed to act. The decision of which controller should be used is based on a handful of variables related to the current state of the system. This scheme can work provided that for each state an adequate/optimal controller can be uniquely determined. When the controllers in the set share the same structure and differ only in some parameters, the problem of controller selection becomes one of parameter selection. These parameters can be computed to be optimal in a, probably small, region of the system's operating space. In this sense, the tuning (parameter selection) takes place locally, giving rise to the denomination of Local Tuning Parameters.

3.2. Partition of Operating Space

The selection of regions where local controllers are defined is not a trivial task in a general case. For the particular case of PCC of an IM drive, expert knowledge suggests using speed ω and load T as scheduling variables. Considering the range of variation of these variables for a particular IM, the operating space can be defined as $\Phi = [0, \overline{\omega}] \times [0, \overline{T}]$ where the over line indicates maximum value.

The partition of Φ can be done in different ways, being the simpler a set of rectangular cells in the form $[h\Delta\omega, (h+1)\Delta\omega] \times [j\Delta T, (j+1)\Delta T]$ obtained considering some increments $\Delta\omega$ and ΔT. For smaller increments, the obtained partition is finer enabling a higher possibility of obtaining an adequate scheduling.

Once the partition has been made, the MBPC λ parameters can be found via simulation for each region as will be shown next.

3.3. Local Tuning

For each cell $\phi_{hj} \in \Phi$ defined as $\phi_{hj} = [h\Delta\omega, (h+1)\Delta\omega] \times [j\Delta T, (j+1)\Delta T]$ the controller parameters are selected as the solutions of Equation (4). The limits can be the same for all operating points or follow some other rule. This is important as the minimization problem of Equation (4) can have no solutions if the limits are too tight. To solve Equation (4), an optimization algorithm linked to a simulation of the drive must be used. The five-phase IM, the VSI and the PCC are simulated using a Runge–Kutta method including the controller as a discrete-time part considering its computing times. The IM parameters are those of the real IM in the experimental setup that are used later for confirmation. A sampling time of $80 \cdot 10^{-6}$ s has been used for the controller. This sampling time is enough for most modern digital signal processors to run the PCC code. Following the idea of locally tuned controllers, the simulation considers operation around each of the considered center of the partition of the operating space. The magnitude of changes must ensure that the operation remains inside the considered cell. The order of events have been found not to alter the results provided that the simulations contain enough data samples (i.e., they are not too short).

4. Results

The constraint satisfaction capability of the proposed controller is assessed using computer simulations and laboratory tests in the experimental setup shown in Figure 2. The equipment used includes a 30-slot symmetrical five-phase induction machine with distributed windings and three pole pairs. The IM is electrically supplied by means of two three-phase two-level inverters (Semikron SKS22F modules), one of which has an unused phase. A DC-link voltage of 300 (V) is applied to both modules. The predictive controller runs on a TMS320F28335 digital signal processor embedded in a MSK28335 Technosoft board with the appropriate digital and analog input/output connections. The rotor mechanical speed is measured using a GHM510296R/2500 digital rotatory encoder. The experimental setup also includes an independently controlled DC machine that is used to produce load torque in the shaft of the IM machine. In this way, different loading conditions can be tested. The electrical parameters (inductances and resistances) have been identified through experimentation, as explained in [23], and are shown in Table 1.

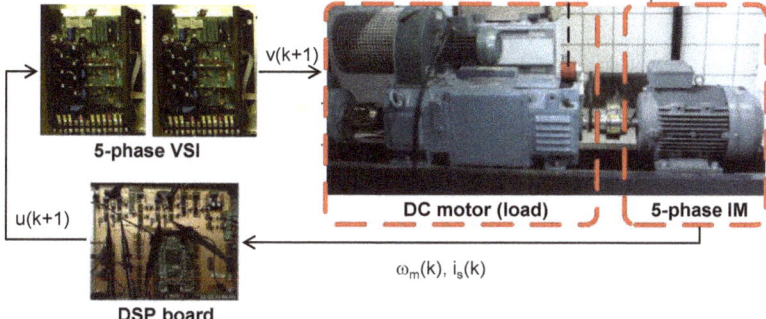

Figure 2. Photographs of the various elements of the experimental setup.

Table 1. Estimated parameters of the IM.

Parameter		Value	Parameter		Value
Stator resistance	R_s (Ω)	19.45	Rotor resistance	R_r (Ω)	6.77
Stator leakage inductance	L_{ls} (mH)	100.7	Rotor leakage inductance	L_{lr} (mH)	38.6
Mutual inductance	L_m (mH)	656.5	Nominal current	I_n (A)	2.5
Mechanical nominal speed	ω_n (rpm)	1000	Nominal torque	T_n (N·m)	4.7

In Figure 3(left), the feasible region for $U_{\alpha\beta} = 0.035$ (A), $U_{sw} = 5500$ Hz is shown for an operating point with nominal speed and load. The solution of Equation (4) is indicated with a times mark ($\times$) corresponding to $\lambda_{xy} = 1$, $\lambda_{sw} = 92 \times 10^{-5}$. It can be seen that the optimal solution is close to the edge of the feasible region as it usually happens in constrained optimization problems. Another example is presented in Figure 3(right) where the operating point is characterized by low speed and load (about 30% of nominal value). Please note that the low speed zone is challenging due to the apparition of large $x - y$ currents [24]. In this case, the solution of Equation (4) takes place for $\lambda_{xy} = 1.64$, $\lambda_{sw} = 51 \times 10^{-5}$. It can be seen that the optimal parameters are quite different for the two operating points considered, even if the admissible limits U are not changed.

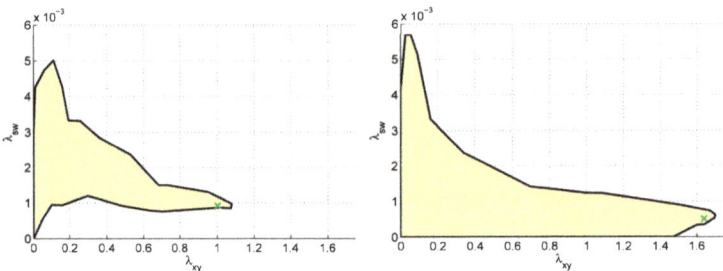

Figure 3. Examples of feasible regions for nominal speed and load (**left**) and for low speed and load (**right**). The optimal solution is shown as a × mark on each region.

The same procedure is repeated for a partition of the operating space, producing an optimal value for the λ parameters that characterizes the optimal controller for each cell. In Table 2, the results are shown for a partition of moderate size (3 × 3). Please note that, for finer partitions, better results can be expected at the cost of more experimentation needed to obtain the local parameters. The acceptable limits are set as in the previous case as $U_{\alpha\beta} = 0.035$ (A), $U_{sw} = 5500$ Hz. The rows and columns in Table 2 are the indices (h, j) that define the cell as $[h\Delta\omega, (h+1)\Delta\omega] \times [j\Delta T, (j+1)\Delta T]$ with $\Delta\omega = 330$ (rmp) and $\Delta T = 1$ (A). The values inside each cell are the pair $\theta = (\lambda_{xy}, \lambda_{sw})$. It can be seen that the optimal values of λ_{xy} lie in the interval [0.15, 1.64] meaning that the higher value is an order of magnitude larger than the lower. Similarly, for λ_{sw}, the interval is $[11, 120] \times 10^{-5}$. In addition, a nonlinear and not obvious relationship between both parameters is appreciable.

Table 2. Optimal θ parameters for $U_{\alpha\beta} = 0.035$ (A), $U_{sw} = 5500$ (Hz).

	1	2	3
1	$(1.64, 51 \times 10^{-5})$	$(1.21, 56 \times 10^{-5})$	$(1.31, 68 \times 10^{-5})$
2	$(1.44, 11 \times 10^{-5})$	$(1.00, 92 \times 10^{-5})$	$(0.68, 120 \times 10^{-5})$
3	$(1.19, 16 \times 10^{-5})$	$(0.17, 86 \times 10^{-5})$	$(0.15, 24 \times 10^{-5})$

Controller Assessment

To assess the proposed scheduled PCC, a comparison with the traditional PCC is made. Several points covering the whole operating space have been considered. For each operating point, the constraints satisfaction is tested by checking the inequalities $E_{\alpha\beta} < U_{\alpha\beta}$ and max $f_{sw} < U_{sw}$. The results are presented with the help of the graph in Figure 4, where red color is used to indicate constraint violation and green for constraint satisfaction. The left graph (Case A) is for $\lambda_{xy} = 0.5$, $\lambda_{sw} = 0$ which is used in a variety of publications [6,23]. The graph on the right (Case B) is obtained for $\lambda_{xy} = 0.14$, $\lambda_{sw} = 60 \times 10^{-5}$. For the proposed strategy of scheduled local parameters, all points satisfy the constraints and thus no graph is needed.

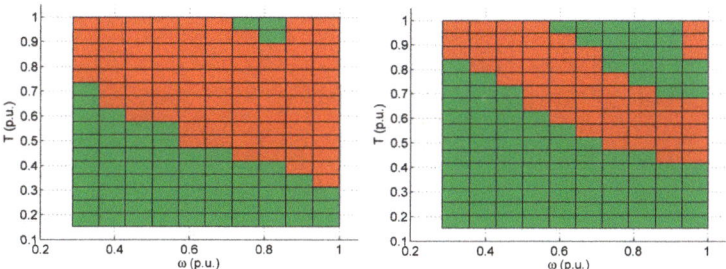

Figure 4. Constraint satisfaction of two PCC with fixed parameters: Case A for $\lambda_{xy} = 0.5$, $\lambda_{sw} = 0$ (**left**) and Case B for $\lambda_{xy} = 0.1$, $\lambda_{sw} = 60 \times 10^{-5}$ (**right**).

The minimization of E_{xy} is now checked. Figure 5 shows the histograms of E_{xy} for the two fixed controllers (Cases A and B above) and for the proposed scheduled PCC using for each operating point the value of θ indicated in Table 2. Please note that in the histogram all points are considered and not just on the green zone (where constraints are satisfied). It can be seen that the proposed controller provides the most adequate distribution of E_{xy} values, being placed at the lower end of the range. This comes in addition to meeting the constraints for all operating points, as already discussed. It is also interesting to note that Case A is better than Case B in terms of E_{xy} but its region of constraint satisfaction is more limited than that of Case B, as previously shown.

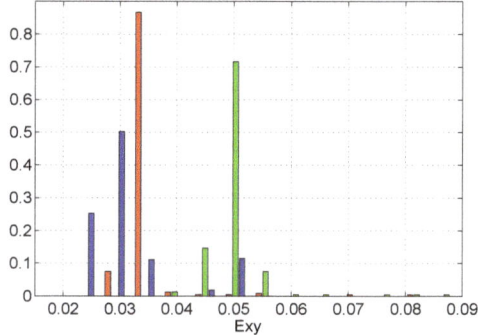

Figure 5. Histograms of E_{xy} for two traditional PCC: Case A (**red**) and Case B (**green**) and for the proposed scheduled PCC (**blue**).

An experimental comparison of the proposed locally tuned controller with a traditional PCC with fixed weighting factors is now presented. Figure 6 shows the trajectories of stator currents in α and x axes (similar results are logically obtained for β and y axes), along with the reference for α currents (for x currents the reference is zero as $x - y$ subspace generates only losses). Two operation points are considered: top row is for nominal speed and low load (Case A) and bottom row for nominal speed and 50% external torque (Case B). The tuning has been performed in this case to achieve tracking error below $U_{\alpha\beta} = 0.18$ (A) and a low commutation rate below $U_{sw} = 4$ kHz. The tuning for the fixed weights PCC is found to be $\lambda_{xy} = 0.5$, $\lambda_{sw} = 1200 \times 10^{-5}$. The locally tuned controller uses $\lambda_{xy} = 0.8$, $\lambda_{sw} = 1000 \times 10^{-5}$ for Operating Point A and $\lambda_{xy} = 0.5$, $\lambda_{sw} = 800 \times 10^{-5}$ for Operating Point B. From the results shown in Figure 6, it is clear that the extra degrees of freedom offered by the local tuning is exploited to obtain better tracking and less $x - y$ content without violating the constraints.

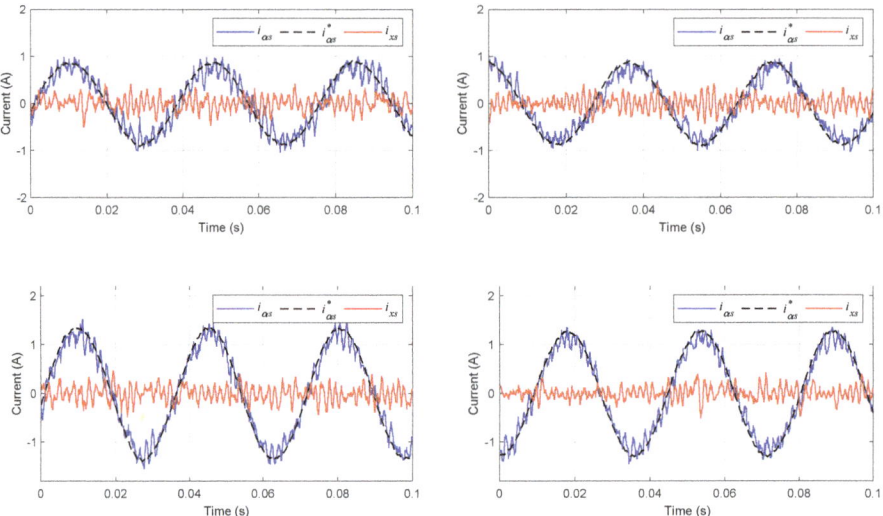

Figure 6. Experimental comparison of the proposed controller (**right**) against a traditional PCC using fixed weighting factors (**left**) at two operating regimes (**top** and **bottom** rows).

5. Conclusions

It has been shown how a modified tuning procedure, mathematically represented by an optimization problem, can solve the constraint handling problem of predictive stator current control for IM thanks to the use of locally tuned controllers. From the experiments, it can be concluded that the constraint feasibility problem has, in general, no solution for standard predictive control. The proposed scheme provides a means of obtaining good tracking performance without violating constraints in switching frequency and at the same time reducing parasitic currents of $x - y$ subspaces.

Author Contributions: Conceptualization, A.K.G. and M.R.A.; methodology, F.B. and M.R.A.; software, A.K.G. and C.M.; validation, A.K.G. and C.M.; formal analysis, A.K.G., M.R.A. and C.M.; investigation, A.K.; resources, F.B.; data curation, A.K.G.; writing—original draft preparation, A.K.G. and M.R.A.; writing—review and editing, A.K., M.R.A. and F.B.; visualization, F.B.; supervision, F.B. and M.R.A.; project administration, A.K. and C.M.; and funding acquisition, M.R.A.

Funding: This research was funded by Ministerio de Ciencia, Innovación y Universidades of Spain grant number RTI2018-101897-B-I00.

Conflicts of Interest: The authors declare no conflict of interest.

References

1. Holmes, D.; Martin, D. Implementation of a direct digital predictive current controller for single and three phase voltage source inverters. In Proceedings of the IAS '96. Conference Record of the 1996 IEEE Industry Applications Conference Thirty-First IAS Annual Meeting, San Diego, CA, USA, 6–10 October 1996; IEEE: Piscataway, NJ, USA, 1996; Volume 2, pp. 906–913.
2. Martin, C.; Barrero, F.; Arahal, M.R.; Duran, M.J. Model-Based Predictive Current Controllers in Multiphase Drives Dealing with Natural Reduction of Harmonic Distortion. *Energies* **2019**, *12*, 1679. [CrossRef]
3. Liu, C.; Luo, Y. Overview of advanced control strategies for electric machines. *Chin. J. Electr. Eng.* **2017**, *3*, 53–61.

4. Tenconi, A.; Rubino, S.; Bojoi, R. Model Predictive Control for Multiphase Motor Drives—A Technology Status Review. In Proceedings of the 2018 International Power Electronics Conference (IPEC-Niigata 2018-ECCE Asia), Niigata, Japan, 20–24 May 2018; IEEE: Piscataway, NJ, USA, 2018; pp. 732–739.
5. Camacho, E.F.; Bordons, C. *Model Predictive Control*; Springer: Berlin/Heidelberg, Germany, 2013.
6. Lim, C.S.; Levi, E.; Jones, M.; Rahim, N.; Hew, W.P. A Comparative Study of Synchronous Current Control Schemes Based on FCS-MPC and PI-PWM for a Two-Motor Three-Phase Drive. *IEEE Trans. Ind. Electron.* **2014**, *61*, 3867–3878. [CrossRef]
7. Arahal, M.R.; Barrero, F.; Durán, M.J.; Ortega, M.G.; Martín, C. Trade-offs analysis in predictive current control of multi-phase induction machines. *Control Eng. Pract.* **2018**, *81*, 105–113. [CrossRef]
8. Mamdouh, M.; Abido, M.; Hamouz, Z. Weighting Factor Selection Techniques for Predictive Torque Control of Induction Motor Drives: A Comparison Study. *Arab. J. Sci. Eng.* **2018**, *43*, 433–445. [CrossRef]
9. Hannan, M.; Ali, J.A.; Mohamed, A.; Hussain, A. Optimization techniques to enhance the performance of induction motor drives: A review. *Renew. Sustain. Energy Rev.* **2018**, *81*, 1611–1626. [CrossRef]
10. Arahal, M.R.; Kowal, A.; Barrero, F.; Castilla, M. Cost Function Optimization for Multi-phase Induction Machines Predictive Control. *Rev. Iberoam. Autom. Inf. Ind.* **2019**, *16*, 48–55. [CrossRef]
11. Khan, H.S.; Aamir, M.; Ali, M.; Waqar, A.; Ali, S.U.; Imtiaz, J. Finite Control Set Model Predictive Control for Parallel Connected Online UPS System under Unbalanced and Nonlinear Loads. *Energies* **2019**, *12*, 581. [CrossRef]
12. Abbaszadeh, A.; Khaburi, D.A.; Kennel, R.; Rodríguez, J. Hybrid exploration state for the simplified finite control set-model predictive control with a deadbeat solution for reducing the current ripple in permanent magnet synchronous motor. *IET Electr. Power Appl.* **2017**, *11*, 823–835. [CrossRef]
13. Vali, M.; Petrovic, V.; Boersma, S.; van Wingerden, J.W.; Pao, L.Y.; Kuhn, M. Adjoint-based model predictive control for optimal energy extraction in waked wind farms. *Control Eng. Pract.* **2019**, *84*, 48–62. [CrossRef]
14. Duran, M.J.; Levi, E.; Barrero, F. Multiphase Electric Drives: Introduction. In *Wiley Encyclopedia of Electrical and Electronics Engineering*; Wiley Online Library: Hoboken, NJ, USA, 2017; pp. 1–26.
15. Kali, Y.; Ayala, M.; Rodas, J.; Saad, M.; Doval-Gandoy, J.; Gregor, R.; Benjelloun, K. Current Control of a Six-Phase Induction Machine Drive Based on Discrete-Time Sliding Mode with Time Delay Estimation. *Energies* **2019**, *12*, 170. [CrossRef]
16. Martín, C.; Arahal, M.R.; Barrero, F.; Durán, M.J. Five-phase induction motor rotor current observer for finite control set model predictive control of stator current. *IEEE Trans. Ind. Electron.* **2016**, *63*, 4527–4538. [CrossRef]
17. Yaramasu, V.; Wu, B. *Model Predictive Control of Wind Energy Conversion Systems*; John Wiley & Sons: Hoboken, NJ, USA, 2016.
18. Rodas, J.; Barrero, F.; Arahal, M.R.; Martín, C.; Gregor, R. Online estimation of rotor variables in predictive current controllers: A case study using five-phase induction machines. *IEEE Trans. Ind. Electron.* **2016**, *63*, 5348–5356. [CrossRef]
19. Gonzalez-Prieto, I.; Duran, M.J.; Aciego, J.J.; Martin, C.; Barrero, F. Model predictive control of six-phase induction motor drives using virtual voltage vectors. *IEEE Trans. Ind. Electron.* **2018**, *65*, 27–37. [CrossRef]
20. Åström, K.J.; Wittenmark, B. On self tuning regulators. *Automatica* **1973**, *9*, 185–199. [CrossRef]
21. Desoer, C. Slowly varying system x = A (t) x. *IEEE Trans. Autom. Control* **1969**, *14*, 780–781. [CrossRef]
22. Hunt, K.J.; Johansen, T.A. Design and analysis of gain-scheduled control using local controller networks. *Int. J. Control* **1997**, *66*, 619–652. [CrossRef]
23. Martín, C.; Bermúdez, M.; Barrero, F.; Arahal, M.R.; Kestelyn, X.; Durán, M.J. Sensitivity of predictive controllers to parameter variation in five-phase induction motor drives. *Control Eng. Pract.* **2017**, *68*, 23–31. [CrossRef]
24. Echeikh, H.; Trabelsi, R.; Iqbal, A.; Bianchi, N.; Mimouni, M.F. Non-linear backstepping control of five-phase IM drive at low speed conditions—Experimental implementation. *ISA Trans.* **2016**, *65*, 244–253. [CrossRef] [PubMed]

© 2019 by the authors. Licensee MDPI, Basel, Switzerland. This article is an open access article distributed under the terms and conditions of the Creative Commons Attribution (CC BY) license (http://creativecommons.org/licenses/by/4.0/).

Article

Min-Max Predictive Control of a Five-Phase Induction Machine

Daniel R. Ramirez [1], Cristina Martin [2,*], Agnieszka Kowal G. [1] and Manuel R. Arahal [1]

1. Systems Engineering and Automation Department, University of Seville, 41092 Seville, Spain; danirr@us.es (D.R.R.); akowal@us.es (A.K.G.); arahal@us.es (M.R.A.)
2. Electronic Engineering Department, University of Seville, 41092 Seville, Spain
* Correspondence: cmartin15@us.es; Tel.: +34-9-5448-7343

Received: 27 August 2019; Accepted: 27 September 2019; Published: 28 September 2019

Abstract: In this paper, a fuzzy-logic based operator is used instead of a traditional cost function for the predictive stator current control of a five-phase induction machine (IM). The min-max operator is explored for the first time as an alternative to the traditional loss function. With this proposal, the selection of voltage vectors does not need weighting factors that are normally used within the loss function and require a cumbersome procedure to tune. In order to cope with conflicting criteria, the proposal uses a decision function that compares predicted errors in the torque producing subspace and in the x-y subspace. Simulations and experimental results are provided, showing how the proposal compares with the traditional method of fixed tuning for predictive stator current control.

Keywords: cost functions; minmax; predictive current control; multi-phase drives

---

1. Introduction

Model Predictive Control (MPC) has been applied to many different types of energy systems [1,2]. In the case of electric machines, the predictive controller can directly command a power converter, typically a Voltage Source Inverter (VSI) yielding a direct digital control scheme [3] that is often referred to as FSMPC and FCSMPC. This scheme has been recently used in many applications, including multi-phase IMs. A particular configuration for IM control is Predictive Stator Current Control (PSCC), which allows us to deal separately with the electro-mechanical aspects of IM control [4].

Multi-phase IMs have lower torque variance, lower DC link current harmonics, and better reliability and power distribution per phase compared with three-phase ones. The most frequent control structure is composed by an inner loop for current control and an outer loop for flux and speed control. Voltage modulation techniques (such as PWM and Space Vector) can be used for current control [5], whereas direct torque control (DTC) uses a switching table to determine the VSI state [6]. In [7], a three-phase to five-phase matrix converter is used to feed a five-phase permanent magnet motor using DTC to eliminate current harmonics. The sensor-less case is explored in [8] for a five-phase interior permanent magnet motor.

The main advantage of predictive schemes is the flexibility to incorporate in the cost function different criteria [6]. In this way different control objectives can be treated with ease. As in other forms of MPC, the strategy in PSCC is to optimize a certain Loss Function (LF) with respect to the control action at each discrete-time period. The LF of PSCC is primarily designed to penalize deviations of stator currents from their references. It is however possible to include additional terms in the LF to penalize (mostly) energy losses. As a result, the design of the LF dictates much of the closed-loop performance of the system. It must be recalled that PSCC has not the ability to simultaneously minimize all of the factors present in the loss function due to the finite number of control moves and the uniform sampling time [9,10]. In fact, one of the reasons for the late popularity of PSCC is its ability to find a compromise solution for colliding objectives present in the LF [11], unlike previous schemes (PWM, DTC) where such criteria are not explicitly considered.

Due to computation time constraints, most PSCC use a prediction horizon of 2 steps and a control horizon of just 1 move. This means that the traditional Weighting Functions (WF) of MPC contain just one value per LF term [12]. For this reason, in the IM control literature, instead of functions these values are referred to as weighting factors. Moreover, in most papers these factors are selected off-line and are kept fixed during operation of the IM. Elimination of the WF has been proposed elsewhere (see [13] for a review) mainly for the Predictive Torque Control (PTC) of conventional (three-phase) IMs.

The proposal of this paper removes the weighting factors by using a min-max decision function where the different sub-spaces, $\alpha - \beta$ and $x - y$, are given a relative importance based solely on their relative values. With this proposal, the selection of voltage vectors do not need weighting factors that require a cumbersome procedure to tune. In order to cope with conflicting criteria the proposal uses a decision function that compares predicted errors in the torque producing subspace and in the $x - y$ subspace. The min-max is a special case of fuzzy logic based functions that have been proposed for use with MPC, where the traditional loss functions are replaced by fuzzy decision functions [14]. To the best of our knowledge the proposal is novel and similar (not exactly the same) schemes have just been applied to torque control in three-phase systems [13]. Other related works do use fuzzy systems to substitute the model or the controller [15–17].

To illustrate the method, in this paper a five-phase drive is considered. This particular system is relevant as the five-phase machine is of interest [6] and the proposed method seeks a trade-off between losses and dynamic performance. Please notice that the strategy is applicable to other types of systems. In the next section, the basic aspects of PSCC are reviewed, introducing the material that will be considered in the proposal for min-max control. Simulation and experimental results are provided for a five-phase IM in Sections 4 and 5, respectively. From these results, some conclusions are derived at the end of the paper.

2. PSCC for Five-Phase IM

The scheme for PSCC for a multi-phase IM contains a digital processor that decides the control action u indicating the state of the VSI to be held for the whole sampling period T_s. Defining the discrete time k such that $t = kT_s$, the actuation signal is denoted as $u(k)$. It is well known that a whole sampling time delay is produced due to computations. To account for this, the controller must select at time k the most appropriate value for $u(k+1)$. The selection is based on minimizing a certain loss function.

Figure 1 shows a diagram where at each discrete-time k the controller computes $u(k+1)$ as

$$u^o(k+1) = \operatorname*{argmin}_{u \in \mathbb{U}} L(k, u), \tag{1}$$

where, $\mathbb{U}$ is the set of all possible control actions (states of the VSI) and L is the loss function. The LF must contain a term penalizing the deviation of predicted stator currents $\hat{i}_s(k+2|k)$ from desired values $r(k+2)$, where predictions depend on past control action $u(k)$ that has been previously set and on the actual control action $u(k+1)$. Predictions are obtained from a model of the system as

$$\hat{y}(k+1|k) = Ay(k) + Bu(k) + G \tag{2}$$
$$\hat{y}(k+2|k) = A\hat{y}(k+1|k) + Bu(k+1), \tag{3}$$

where matrices A and B and vector G are obtained from time-discretization of the systems' dynamic equations (see [4] for details).

The necessary penalization of the deviation of $\hat{i}_s(k+2|k)$ from $r(k+2)$ is actually done in different planes arising from Clarke's transformation. By means of this, the stator currents are mapped into an

$\alpha - \beta$ plane and several $x - y$ and z planes. For the five-phase IM that will be used as a case study, just $\alpha - \beta$ and $x - y$ axes need to be used [18]. With these considerations, the LF can now be defined as

$$L(k, u) = \hat{e}_{\alpha\beta}^2(k, u) + \lambda \cdot \hat{e}_{xy}^2(k, u), \tag{4}$$

where the predicted errors are computed as $\hat{e}_{\alpha\beta}(k, u) = \|r_{\alpha,\beta}(k+2) - \hat{i}_{s\alpha,\beta}(k+2|k, u)\|$, $\hat{e}_{xy}(k, u) = \|\hat{i}_{sx,y}(k+2|k, u)\|$. The reference value for the $x - y$ plane is zero, as it is the normal case. The $\alpha - \beta$ reference is given by $r_\alpha(k) = I \sin 2\pi f_e k T_s$, $r_\beta(k) = I \cos 2\pi f_e k T_s$ where f_e is the electrical frequency determined by the mechanical speed and I is the amplitude that depends on the mechanical load. Both quantities are supplied to the PSCC by the higher level controller responsible for tracking mechanical variables (speed, torque, position), depending on the application.

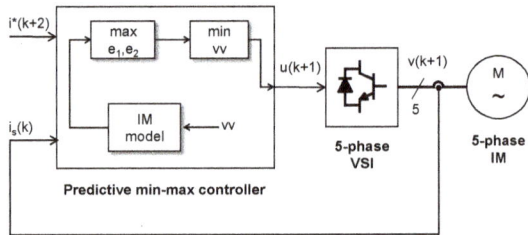

Figure 1. Block diagram of min-max Predictive Stator Current Control (PSCC) for a five-phase induction machine (IM) driven by an Voltage Source Inverter (VSI).

The weighting factor λ_{xy} provides the relative importance of $x - y$ plane regulation over $\alpha - \beta$ tracking. The $\alpha - \beta$ plane is related to power conversion and the $x - y$ to losses. In the traditional PSCC, λ_{xy} is treated as a parameter of the controller, selected off-line and kept constant during operation of the drive. The PSCC then uses (1) to produce $u(k+1)$, which is sent to the VSI and kept for the whole sampling period. This is repeated a new the next sampling period, using the receding horizon strategy [12]. Please note that, as an alternative, virtual voltage vectors have been proposed to produce a sort of modulation producing an average value for $x - y$ voltages of zero [19].

The performance of PSCC is in most cases presented using the tracking error as a figure of merit. For a generic $l - m$ plane, the tracking error is defined as $e_{l,m} = \|r_{l,m}(k) - i_{sl,m}(k)\|$, where (in multi-phase IM), the pair (l, m) usually takes the values (α, β) and $(x - y)$. With a sufficiently accurate model ([20,21]), the control objectives (current tracking in $\alpha - \beta$ and $x - y$ planes) are achieved to some degree. Depending on the application some additional criteria are also reported. Regarding controller tuning, the only parameters needed appear in the IM model (found via identification [21]) and weighting factors, usually computed off-line [22]. The tuning procedure is cumbersome since there is not a direct relationship between figures of merit and WF values; moreover, the WF that yield a particular behavior might change with the operating point [23]. In the next section, the LF is replaced by a fuzzy-logic-based function containing no weighting factors.

3. Min-Max Predictive Stator Current Control

The loss function with weighting factors presented above is not the only way to overcome the problems associated with the multi-objective nature of the selection of the control action. The use of a loss function derived from a fuzzy-logic approach is one possibility to avoid the need of weighting factors tuning eliminating a cumbersome trial and error procedure [23]. In the present case, such function should be chosen according to the criteria of balancing tracking in the $\alpha - \beta$ and $x - y$ sub-spaces to maintain performance with diminished losses.

Several types of fuzzy-logic operators can be use to aggregate the terms in (4). Control objectives are considered as fuzzy goals each expressed as a membership function. The aggregation of membership functions (as considered in the realm of fuzzy logic) allows for the simultaneous consideration of more than one objective in control terms. The set of all used aggregation operators and membership functions allows to treat a linguistic description of objectives in a mathematical way [24]. The minimum operator has been proposed for different applications. However it does not allow to balance different objectives to find a trade-off solution. The product t-norm [25] allows to achieve a trade-off solution, however the importance of different criteria must be equal, otherwise a method to attribute relative importance is needed. This leads to the use of weighting factors and/or parametric t-norms [26]. An important aspect for PSCC is that the computation time is limited to a few microseconds (typically between 40 and 100 µs). This limits the complexity of the loss function to be used since it must be used repeatedly in the optimization phase.

In this paper, the min-max operator is proposed to be used as an alternative to the traditional loss function. The min-max operator is derived from the Minimax decision rule of game theory [27]. It has been used for minimizing the expected loss for a worst case scenario. It can be used within fuzzy logic decision-making schemes and has also been used in MPC [28]. One problem to be solved is the computationally intensive nature of the Minimax rule for MPC [29]. In this particular case, the reduced control horizon allows for a realization in real-time as the evaluation of the min-max loss function is not more demanding than the traditional one. Incorporating the min-max idea, the resulting loss function takes the form

$$L(k,u) = \min_{v} \max\{\hat{e}_{\alpha\beta}(k,u_v), \hat{e}_{xy}(k,u_v)\}, \quad (5)$$

where v is an index defining the VSI state (e.g., for a five-phase VSI $v \in \{1,31\}$). The rationale for this choice is as follows, at any given instant k, the different control actions that the multi-phase VSI can produce u_v are considered. For each one, the predicted errors in the $\alpha - \beta$ and $x - y$ planes are computed and the largest value is selected. The control action to be applied at $k+1$ is selected as the one minimizing the selected maximal errors. In this way extreme values of errors for either plane are avoided. It is important to remark that expression (5) contains no adjustable parameters, yet the selection of the control action is driven by both terms $\alpha - \beta$ and $x - y$ predicted errors. However, the relative importance given to each sub-space is not fixed as the selection is made based on the relative values of errors.

4. Simulations

The proposed controller is tested against a traditional FCSMPC with fixed weighting factors in simulation. The IM has been simulated using the Runge-Kutta method. The controller is also incorporated in the simulation as a discrete-time subsystem. To add more realism to the simulations the effect of the one-sampling time delay is also simulated. A sampling time of 80 (µs) is used for the discrete-time part. Please note that this value is within the range usually found for PSCC and can be obtained by a variety of modern digital signal processors. The IM used in simulation has the electrical parameters shown in Table 1 which corresponds to the laboratory setup that will be used later in real experiments.

Table 1. Estimated parameters of the IM.

Parameter		Value	Parameter		Value
Stator resistance	R_s (Ω)	19.45	Rotor resistance	R_r (Ω)	6.77
Stator leakage inductance	L_{ls} (mH)	100.7	Rotor leakage inductance	L_{lr} (mH)	38.6
Mutual inductance	L_m (mH)	656.5	Nominal current	I_n (A)	2.5
Mechanical nominal speed	ω_n (rpm)	1000	Nominal torque	T_n (N·m)	4.7

The simulations will compare the proposal against a traditional FCSMPC with two different tunings. Two operating regimes are considered with nominal speed and 0 external load (labelled as case S1); and nominal speed at 70% load torque (case S2). The tuning for the traditional controller are $\lambda_{xy} = 0.5$ which is a common choice that appears in many papers and aims at a similar penalization for $\alpha - \beta$ and $x - y$ control errors and $\lambda_{xy} = 0.1$ which has also been proposed in some papers and that seeks a better $\alpha - \beta$ tracking at the expense of some $x - y$ content. Table 2 shows the RMS control error for $\alpha - \beta$ and $x - y$ subspaces for the considered controllers and operating regimes. In said Table the entry PCλ05 corresponds to the traditional FCSMPC with $\lambda_{xy} = 0.5$, PCλ01 corresponds to the traditional FCSMPC with $\lambda_{xy} = 0.1$, and PCminmax corresponds to the proposed controller using a min-max loss function. It is interesting to see that the proposed controller provides low values for the tracking errors in both sub-spaces despite the fact that no tuning has been used. The traditional scheme however can be used to put more or less emphasis on $\alpha - \beta$ tracking versus $x - y$ regulation. This degree of freedom is somehow hindered by the fact that different operating regimes would need a different tuning as exposed in previous works (see [11,23]). The trajectories shown in Figure 2 allow for further comparison of the proposed controller against the traditional FCSMPC with $\lambda_{xy} = 0.5$ for operating regimes S1 and S2 previously considered. The advantages of the proposal are apparent as a more accurate tracking is achieved in both cases.

Table 2. Simulation results for the traditional and proposed predictive controllers in terms of tracking errors.

Case	PCλ05	PCλ01	PCmin-Max
S1	$e_{\alpha-\beta} = 0.0542$ (A)	$e_{\alpha-\beta} = 0.0530$ (A)	$e_{\alpha-\beta} = 0.0531$ (A)
	$e_{x-y} = 0.1221$ (A)	$e_{x-y} = 0.1417$ (A)	$e_{x-y} = 0.1109$ (A)
S2	$e_{\alpha-\beta} = 0.1821$ (A)	$e_{\alpha-\beta} = 0.1117$ (A)	$e_{\alpha-\beta} = 0.1810$ (A)
	$e_{x-y} = 0.0984$ (A)	$e_{x-y} = 0.1098$ (A)	$e_{x-y} = 0.1001$ (A)

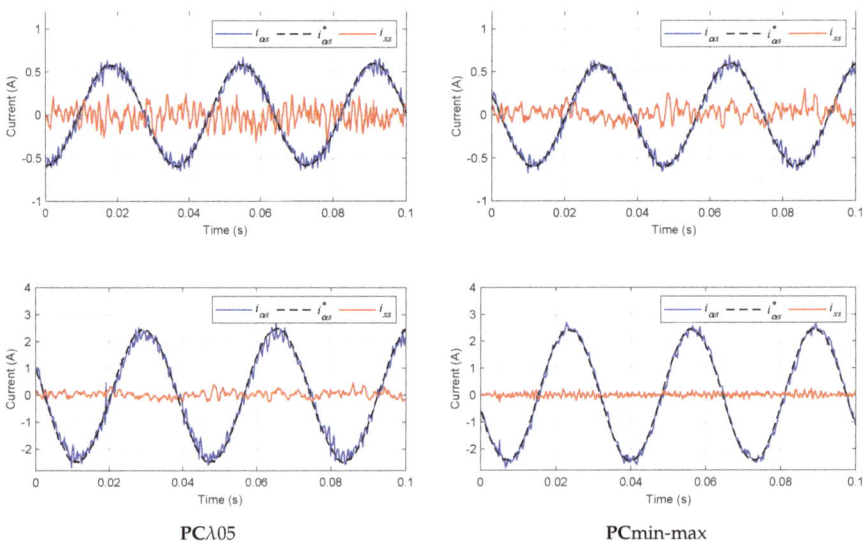

Figure 2. Simulation comparison of the proposed controller (**right**) against a traditional FCSMPC (**left**) at operating regimes S1 and S2 (**top** and **bottom** rows).

5. Experimental Results

The test rig for experimentation is schematically shown in Figure 3, it contains a 30-slot symmetrical five-phase IM with three pole pairs made up by distributed windings. The five-phase two-levels VSI is made of two three-phase Semikron SKS22F modules. The DC-link voltage used is $V_{DC} = 300$ (V). The controllers run on the TMS320F28335 DSP included in a board (MSK28335 Technosoft) which interfaces with the GHM510296R/2500 rotatory encoder. A DC machine is used to emulate the desired load torque in the shaft for the experiments. The IM has the electrical parameters already shown in Table 1.

As in the simulation section the proposed predictive controller with min-max loss function will be compared against a traditional FCSMPC with two different tunings ($\lambda_{xy} = 0.5$ and $\lambda_{xy} = 0.1$). The operating regimes are nominal speed with 4 % load (case E1) and nominal speed with 50 % load torque (case E2). Table 3 shows the RMS control errors. Again, the proposed controller provides low values for the tracking errors in both sub-spaces. The trajectories shown in Figure 4 allow for further comparison of the proposed controller against the traditional FCSMPC with $\lambda_{xy} = 0.5$ for operating points E1 and E2.

The results match very much those already seen in simulation. Please note that a reduction in tracking errors is observed despite the fact that the tuning employed for the traditional FCSMPC is a standard one, found after intensive experimentation (see [22]). In the case of the proposed min-max loss function, the controller aims at minimizing the largest of errors in either sub-space. This results in trajectories that are smoother as extreme deviations are avoided.

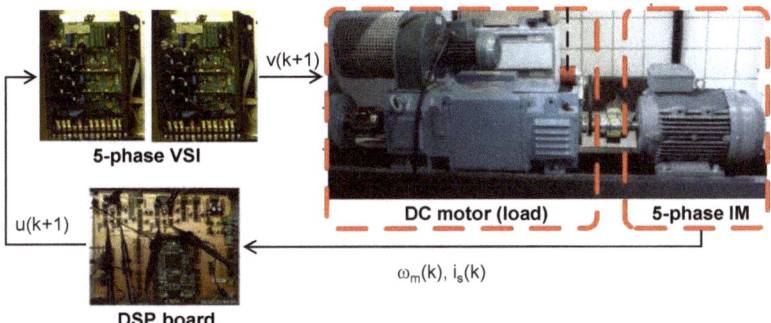

Figure 3. Photographs of the various elements of the experimental setup.

Table 3. Experimental results for the traditional and proposed predictive controllers in terms of tracking errors.

Case	PCλ05	PCλ01	PCmin-Max
E1	$e_{\alpha-\beta} = 0.1416$ (A)	$e_{\alpha-\beta} = 0.1043$ (A)	$e_{\alpha-\beta} = 0.1291$ (A)
	$e_{x-y} = 0.1229$ (A)	$e_{x-y} = 0.1344$ (A)	$e_{x-y} = 0.0983$ (A)
E2	$e_{\alpha-\beta} = 0.1502$ (A)	$e_{\alpha-\beta} = 0.1117$ (A)	$e_{\alpha-\beta} = 0.1411$ (A)
	$e_{x-y} = 0.1185$ (A)	$e_{x-y} = 0.1273$ (A)	$e_{x-y} = 0.1002$ (A)

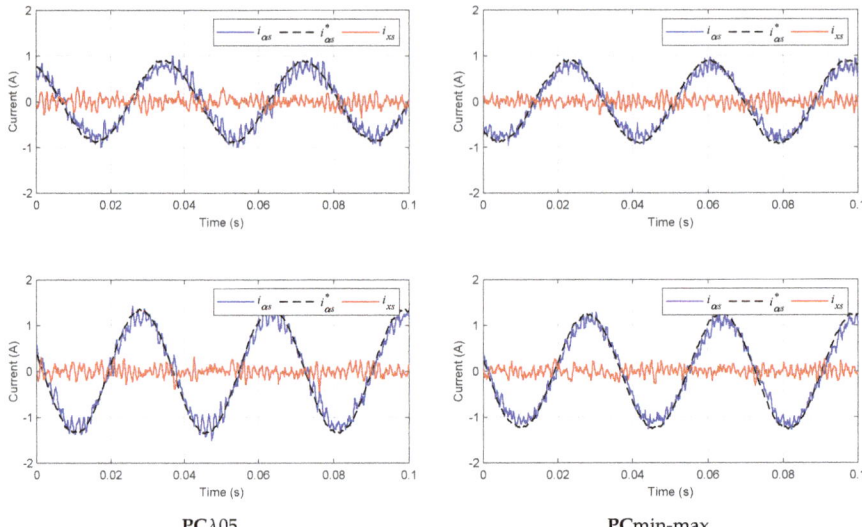

Figure 4. Experimental comparison of the proposed controller (**right**) against a traditional FCSMPC (**left**) at two operating regimes (**top** and **bottom** rows).

A final set of experiments is performed to show the performance for various speeds. Table 4 shows the RMS control error for $\alpha - \beta$ and $x - y$ sub-spaces for the proposed min-max PSCC for different speeds and two loads ($T_L = 4$ (%) and $T_L = 50$ (%)). Again, it is worth remarking the balanced nature of the results for different operating regimes.

Table 4. Experimental results for the proposed min-max PSCC for different speeds and loads.

ω_m (rpm)	$T_L = 4$ (%)	$T_L = 50$ (%)
200	$e_{\alpha-\beta} = 0.1017$ (A)	$e_{\alpha-\beta} = 0.1172$ (A)
	$e_{x-y} = 0.0954$ (A)	$e_{x-y} = 0.1040$ (A)
500	$e_{\alpha-\beta} = 0.1291$ (A)	$e_{\alpha-\beta} = 0.1411$ (A)
	$e_{x-y} = 0.0983$ (A)	$e_{x-y} = 0.1002$ (A)
700	$e_{\alpha-\beta} = 0.1558$ (A)	$e_{\alpha-\beta} = 0.1757$ (A)
	$e_{x-y} = 0.1053$ (A)	$e_{x-y} = 0.1072$ (A)

6. Conclusions

It has been shown that a fuzzy decision making scheme can be used for predictive stator current control of multi-phase IM. Similar ideas have been proposed for predictive torque control of conventional (three-phase) IM and PMSM; they have been refined here to provide a means to cope with specific aspects of multi-phase IMs. The computational requirements are low, allowing for its use in real time, as is demonstrated by the experiments using a standard value for the sampling time. It is interesting to see that the proposed controller provides low values for the tracking errors in torque producing ($\alpha - \beta$) and loss producing ($x - y$) sub-spaces, despite the fact that no tuning is needed.

A good agreement between experimental and simulation results has been found. While watching the results, one must bear in mind that the instantaneous (discrete-time wise) minimization of a loss function does not guarantee a certain result in terms of the trajectories obtained due to the effect of the receding horizon in predictive control. In the case of the proposed min-max loss function, the controller aims at minimizing the largest of errors in either sub-space. This results in trajectories that are observed in the results to be smoother as extreme deviations (in either sub-space) are avoided.

Author Contributions: Conceptualization, D.R.R., C.M. and M.R.A.; methodology, D.R.R. and M.R.A.; software, D.R.R., A.K.G. and C.M.; validation, A.K.G. and C.M.; formal analysis, D.R.R., M.R.A. and C.M.; investigation, D.R.R., C.M. and M.R.A.; resources, D.R.R. and C.M.; data curation, C.M. and A.K.G.; writing—original draft preparation, D.R.R., C.M., A.K.G. and M.R.A.; writing—review and editing, D.R.R., A.K.G. and M.R.A.; visualization, D.R.R.; supervision, D.R.R. and M.R.A.; project administration, A.K.G. and C.M.; funding acquisition, D.R.R.

Funding: This work has received support by MINECO-Spain, FEDER Funds and University of Seville under grants DPI2016-76493-C3-1-R and 2014/425, respectively.

Conflicts of Interest: The authors declare no conflict of interest.

References

1. Alamin, Y.; Castilla, M.; Álvarez, J.; Ruano, A. An economic model-based predictive control to manage the users' thermal comfort in a building. *Energies* **2017**, *10*, 321. [CrossRef]
2. Camacho, E.; Gallego, A.; Sanchez, A.; Berenguel, M. Incremental state-space model predictive control of a Fresnel solar collector field. *Energies* **2019**, *12*, 3. [CrossRef]
3. Holmes, D.; Martin, D. Implementation of a direct digital predictive current controller for single and three phase voltage source inverters. In Proceedings of the 1996 IEEE Industry Applications Conference Thirty-First IAS Annual Meeting, San Diego, CA, USA, 6–10 October 1996; Volume 2, pp. 906–913.
4. Arahal, M.R.; Barrero, F.; Toral, S.; Durán, M.J.; Gregor, R. Multi-phase current control using finite-state model-predictive control. *Control Eng. Pract.* **2009**, *17*, 579–587. [CrossRef]
5. Prieto, J.; Jones, M.; Barrero, F.; Levi, E.; Toral, S. Comparative analysis of discontinuous and continuous PWM techniques in VSI-fed five-phase induction motor. *IEEE Trans. Ind. Electron.* **2011**, *58*, 5324–5335. [CrossRef]
6. Duran, M.J.; Levi, E.; Barrero, F. *Multiphase Electric Drives: Introduction*; Wiley Encyclopedia of Electrical and Electronics Engineering: New Jersey, USA, 2017; pp. 1–26.
7. Yousefi-Talouki, A.; Gholamian, S.A.; Yousefi-Talouki, M.; Ilka, R.; Radan, A. Harmonic elimination in switching table-based direct torque control of five-phase PMSM using matrix converter. In Proceedings of the 2012 IEEE Symposium on Humanities, Science and Engineering Research, Kuala Lumpur, Malaysia, 24–27 June 2012; pp. 777–782.
8. Parsa, L.; Toliyat, H.A. Sensorless direct torque control of five-phase interior permanent-magnet motor drives. *IEEE Trans. Ind. Appl.* **2007**, *43*, 952–959. [CrossRef]
9. Arahal, M.R.; Barrero, F.; Ortega, M.G.; Martin, C. Harmonic analysis of direct digital control of voltage inverters. *Math. Comput. Simul.* **2016**, *130*, 155–166. [CrossRef]
10. Xu, Y.; Shi, T.; Yan, Y.; Gu, X. Dual-Vector predictive torque control of permanent magnet synchronous motors based on a candidate vector table. *Energies* **2019**, *12*, 163. [CrossRef]
11. Arahal, M.R.; Barrero, F.; Durán, M.J.; Ortega, M.G.; Martín, C. Trade-offs analysis in predictive current control of multi-phase induction machines. *Control Eng. Pract.* **2018**, *81*, 105–113. [CrossRef]
12. Camacho, E.F.; Bordons, C. *Model Predictive Control*; Springer: Berlin, Germany, 2013.
13. Mamdouh, M.; Abido, M.; Hamouz, Z. Weighting Factor Selection Techniques for Predictive Torque Control of Induction Motor Drives: A Comparison Study. *Arab. J. Sci. Eng.* **2018**, *43*, 433–445. [CrossRef]
14. Yen, J.; Langari, R.; Zadeh, L.A. *Industrial Applications of Fuzzy Logic and Intelligent Systems*; IEEE Press: Piscataway, NJ, USA, 1995.
15. Uddin, M.N.; Radwan, T.S.; Rahman, M.A. Performances of fuzzy-logic-based indirect vector control for induction motor drive. *IEEE Trans. Ind. Appl.* **2002**, *38*, 1219–1225. [CrossRef]
16. Neema, D.; Patel, R.; Thoke, A. Speed control of induction motor using fuzzy rule base. *Int. J. Comput. Appl.* **2011**, *33*, 21–29.
17. El-Barbary, Z. Fuzzy logic based controller for five-phase induction motor drive system. *Alex. Eng. J.* **2012**, *51*, 263–268. [CrossRef]

18. Martín, C.; Arahal, M.R.; Barrero, F.; Durán, M.J. Five-phase induction motor rotor current observer for finite control set model predictive control of stator current. *IEEE Trans. Ind. Electron.* **2016**, *63*, 4527–4538. [CrossRef]
19. Entrambasaguas, P.G.; Prieto, I.G.; Martínez, M.J.D.; Guzmán, M.B.; García, F.J.B. Vectores Virtuales de Tensión en Control Directo de Par para una Máquina de Inducción de Seis Fases. *Revis. Iberoam. Autom. Inf. Ind.* **2018**, *15*, 277–285. [CrossRef]
20. Rodas, J.; Barrero, F.; Arahal, M.R.; Martín, C.; Gregor, R. Online estimation of rotor variables in predictive current controllers: A case study using five-phase induction machines. *IEEE Trans. Ind. Electron.* **2016**, *63*, 5348–5356. [CrossRef]
21. Martín, C.; Bermúdez, M.; Barrero, F.; Arahal, M.R.; Kestelyn, X.; Durán, M.J. Sensitivity of predictive controllers to parameter variation in five-phase induction motor drives. *Control Eng. Pract.* **2017**, *68*, 23–31. [CrossRef]
22. Lim, C.S.; Levi, E.; Jones, M.; Rahim, N.; Hew, W.P. A Comparative Study of Synchronous Current Control Schemes Based on FCS-MPC and PI-PWM for a Two-Motor Three-Phase Drive. *IEEE Trans. Ind. Electron.* **2014**, *61*, 3867–3878. [CrossRef]
23. Arahal, M.R.; Kowal, A.; Barrero, F.; Castilla, M. Cost Function Optimization for Multi-phase Induction Machines Predictive Control. *Rev. Iberoam. Autom. Inf. Ind.* **2019**, *16*, 48–55. [CrossRef]
24. Aracil, J.; Gordillo, F.; Alamo, T. Global Stability Analysis of Second-Order Fuzzy Systems. In *Advances in Fuzzy Control*; Springer: Berlin, Germany, 1998; pp. 11–31.
25. Wang, L.X. Stable adaptive fuzzy control of nonlinear systems. *IEEE Trans. Fuzzy Syst.* **1993**, *1*, 146–155. [CrossRef]
26. Yager, R.R. On the inclusion of importances in multi-criteria decision making in the fuzzy set framework. *Int. J. Expert Syst. Res. Appl.* **1992**, *5*, 211–228.
27. Savage, L.J. The theory of statistical decision. *J. Am. Stat. Assoc.* **1951**, *46*, 55–67. [CrossRef]
28. Ramırez, D.; Camacho, E.; Arahal, M. Implementation of min–max MPC using hinging hyperplanes. Application to a heat exchanger. *Control Eng. Pract.* **2004**, *12*, 1197–1205. [CrossRef]
29. Ramírez, D.R.; Alamo, T.; Camacho, E.F. Computational burden reduction in min–max MPC. *J. Frankl. Inst.* **2011**, *348*, 2430–2447.

© 2019 by the authors. Licensee MDPI, Basel, Switzerland. This article is an open access article distributed under the terms and conditions of the Creative Commons Attribution (CC BY) license (http://creativecommons.org/licenses/by/4.0/).

Article

Assessment of a Universal Reconfiguration-less Control Approach in Open-Phase Fault Operation for Multiphase Drives

Federico Barrero [1,*], Mario Bermudez [2], Mario J. Duran [3], Pedro Salas [3] and Ignacio Gonzalez-Prieto [3]

1. Electronic Engineering Department, University of Seville, 41092 Sevilla, Spain
2. Electrical and Thermal Engineering Department, University of Huelva, 21007 Huelva, Spain; mariobermg@gmail.com
3. Electrical Engineering Department, University of Málaga, 29071 Málaga, Spain; mjduran@uma.es (M.J.D.); psbiedma@uma.es (P.S.), igp@uma.es (I.G.-P.)
* Correspondence: fbarrero@us.es; Tel.: +34-954-48-13-04

Received: 7 November 2019; Accepted: 5 December 2019; Published: 10 December 2019

Abstract: Multiphase drives have been important in particular industry applications where reliability is a desired goal. The main reason for this is their inherent fault tolerance. Different nonlinear controllers that do not include modulation stages, like direct torque control (DTC) or model-based predictive control (MPC), have been used in recent times to govern these complex systems, including mandatory control reconfiguration to guarantee the fault tolerance characteristic. A new reconfiguration-less approach based on virtual voltage vectors (VVs) was recently proposed for MPC, providing a natural healthy and faulty closed-loop regulation of a particular asymmetrical six-phase drive. This work validates the interest in the reconfiguration-less approach for direct controllers and multiphase drives.

Keywords: multiphase induction motor drives; natural fault tolerance; virtual voltage vectors

1. Introduction

The use of variable-speed drives has grown in the last few decades because of microprocessor and power converter development, with an expectation that 80% of all the produced energy will be used in locomotive traction, electric ship propulsion, more-electric aircraft and renewable energy applications. Although three-phase machines are the common trend, the interest of the research community has recently focused on machines with more than three phases, named multiphase machines, due to their advantages in terms of reliability and postfault usage [1]. This is the case of safety applications, where the fault-tolerant ability of multiphase drives has attracted the interest of the scientific community. Having more than three phases allows the faulty operation of the drive under specific voltage, current and/or power limits, which makes multiphase drives an interesting solution in critical industry applications, i.e., offshore wind generators, more-electric aircraft and field or traction applications, e.g., the Royal Navy Type-45 destroyer [2]. Thus, multiphase drives can still be operated without the need for heavy topological changes in the power converter, even if a phase is missing, provided that the number of remaining phases is equal to or greater than three [3–5].

Most of the recent multiphase drive research has been focused on the extension of the control methods usually applied in conventional three-phase drives, along with their enhancement to provide the best performance of the system [6,7]. Compared to classical three-phase drives, multiphase ones reduce the electrical stress in drives and power electronic components, since they can manage more power with lower torque pulsation and current harmonic contents [8]. In this context, the most common control strategy is the field-oriented control (FOC) technique, based on linear cascaded control loops and assisted by coordinate transformations and modulation stages. The multiphase machine is decomposed into multiple orthogonal d–q subspaces (fundamental and harmonics components), with each set of d–q variables being independently controlled. The reference voltages are determined by proportional–integral (PI) controllers, which are inputs for the modulation stage to generate pulse-width-modulation (PWM) switching signals to a power converter, usually a voltage-source converter (VSC). Alternatives to this technique are the 'direct control' methods, including direct torque and model-based predictive controllers (DTC and MPC, respectively). They directly switch the VSC state, avoiding the PWM stage and forcing the controlled variables to rapidly track the reference, while achieving normal operation of the drive [1,2,5,8].

Since the healthy operation of the multiphase drive is a significant and complex issue, important and recent research papers also analyze the implications of faulty operation from the control perspective, where FOC, DTC and MPC methods have been studied. It is interesting to note that control techniques based on 'nonlinear' controllers (DTC and MPC methods or 'direct' controllers) appear as promising control alternatives due to their flexibility and simple formulation, competing with FOC techniques for leading the control solution in the field of multiphase machines and drives [5,8]. The term 'fault tolerance' has broad application since the fault can occur in many elements of the system, including VSC and machine faults that lead to short-circuit (i.e., phase [9], VSC switch [10], interturn [11]) or open-circuit (i.e., phase [12], VSC switch [13], or line [14]) faults. This work analyzes the field of multiphase drives and their use in open-phase fault operations, where the recent definition of a reconfiguration-less MPC controller for asymmetrical six-phase drives, based on virtual voltage vectors (VVs) and useful in healthy and faulty operation, seems to be an interesting advance for using direct controllers [15,16]. This work goes beyond [15,16], extending and experimentally validating the idea to different direct controllers and multiphase drives.

2. Basis of Natural Fault-Tolerant Controllers Using MPC

The research activity in the application of MPC techniques in the field of multiphase drives has recently given rise to numerous control approaches, the most popular being the finite control set MPC (FCS-MPC) method due to the limited number of possible switching states of the power converter. The most common use is the result of its combination with FOC methods, with the outer speed control loop being maintained while the inner current regulators are substituted by FCS-MPC controllers. This control scheme is depicted in Figure 1, where a distributed-winding five-phase Induction Motor (IM) drive is used as a case example.

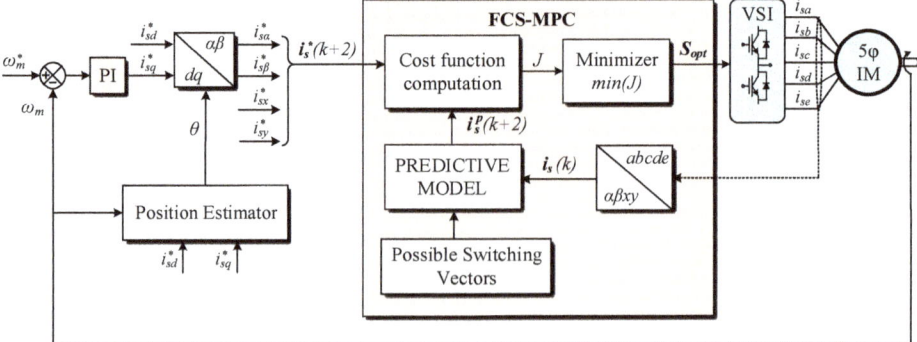

Figure 1. Finite control set model-based predictive control (FCS-MPC) control scheme for a five-phase IM drive fed using a voltage-source inverter (VSI).

The fault-tolerant capability of multiphase drives was first studied in [17], where it was shown that n-phase machines can operate after one or several fault occurrences, as long as the number of healthy phases remains greater than or equal to three and at the expense of a reduction in the torque production. The FCS-MPC control system technology finds itself currently at a paradigm-changing tipping point, where emerging applications are under development. In this context, FCS-MPC has proven to be a promising alternative in the fault-tolerance control of the drive, where it was recently shown [16] that it is possible to skip any control reconfiguration if virtual voltage vectors (VVs) are used. The main idea, shown in Figure 2 for our case example, is to substitute the available voltage vectors with new virtual voltage vectors. Large and medium voltage vectors, which are aligned in the α–β plane with opposite directions in the x–y plane, are combined to provide zero average voltage in the x–y subspace, so harmonic currents in the x–y subspace are reduced. This has been termed natural fault-tolerance capability, and it has been verified so far by using FCS-MPC strategies and six-phase drives [16], where the use of VVs introduces an open-loop control of the stator current in the x–y plane, avoiding the main problems of full-order closed-loop controllers when the open-phase fault occurs and the current cannot flow through a damaged phase, e.g., searching for incompatible control goals and voltage vectors that are no longer available.

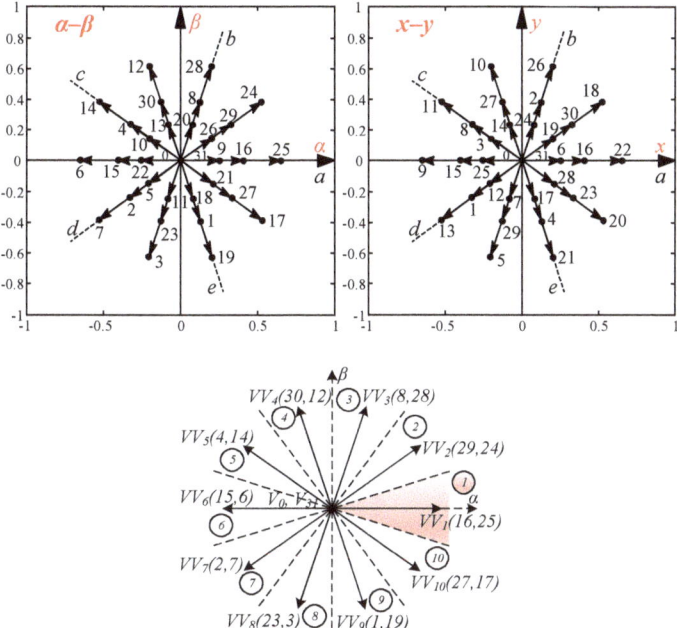

Figure 2. Voltage vectors (VVs) in a five-phase IM drive: (**upper plots**) α–β and x–y subspaces; (**lower plot**) defined VVs.

3. Extension of the Reconfiguration-less Approach to the DTC Case

DTC-based controllers have been also recently proposed and studied in the multiphase drive field for healthy and faulty modes of operation (Figure 3) [18–21]. The basis is to select a stator voltage vector, according to Table 1 and the VV concept, using hysteresis-based controllers for every control period, to obtain reference torque and stator flux tracking. For such purpose, the model of the machine is used to estimate the stator flux and the electromagnetic torque for the DTC controller [22]. In fact, the VVs concept was originally introduced as a way to extend DTC to the multiphase field, maximizing the torque production while minimizing harmonic components in the x–y plane. Hence, the use of VVs in DTC also avoids the problems with full-order closed-loop controllers when open-phase faults occur. This work demonstrates that DTC strategies using VVs can also provide the natural fault-tolerant capability, allowing a ripple-free postfault operation with no reconfiguration of the control strategy. Then, opposite to [21], lookup tables (see Table 1) are not reconstructed after a fault occurrence, and the same control scheme and VVs are used in the pre- and postfault situations.

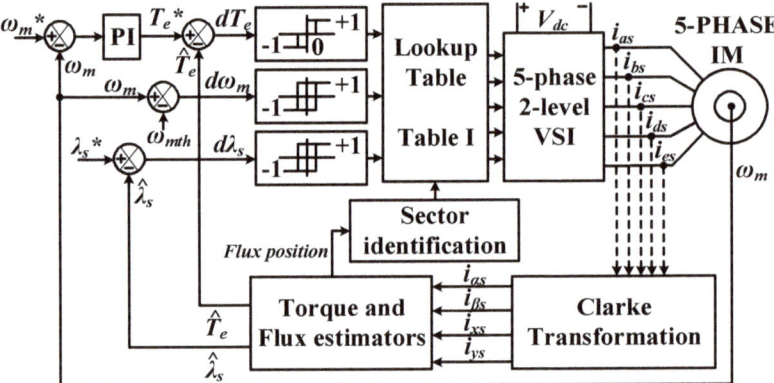

Figure 3. Control scheme of a five-phase IM drive using the direct torque controller (DTC) technique.

Table 1. Lookup table for the DTC controller in healthy operation.

$d\lambda_s$	dT_e	$d\omega_m$	Position of Stator Flux (Sector)									
			1	2	3	4	5	6	7	8	9	10
+1	+1	+1	VV_3	VV_4	VV_5	VV_6	VV_7	VV_8	VV_9	VV_{10}	VV_1	VV_2
		−1	VV_2	VV_3	VV_4	VV_5	VV_6	VV_7	VV_8	VV_9	VV_{10}	VV_1
	−1	+1	VV_9	VV_{10}	VV_1	VV_2	VV_3	VV_4	VV_5	VV_6	VV_7	VV_8
		−1	VV_{10}	VV_1	VV_2	VV_3	VV_4	VV_5	VV_6	VV_7	VV_8	VV_9
	0	+1	V_0	V_{31}	V_0	V_{31}	V_0	V_{31}	V_0	V_{31}	V_0	V_{31}
		−1	V_0	V_{31}	V_0	V_{31}	V_0	V_{31}	V_0	V_{31}	V_0	V_{31}
−1	+1	+1	VV_4	VV_5	VV_6	VV_7	VV_8	VV_9	VV_{10}	VV_1	VV_2	VV_3
		−1	VV_5	VV_6	VV_7	VV_8	VV_9	VV_{10}	VV_1	VV_2	VV_3	VV_4
	−1	+1	VV_8	VV_9	VV_{10}	VV_1	VV_2	VV_3	VV_4	VV_5	VV_6	VV_7
		−1	VV_7	VV_8	VV_9	VV_{10}	VV_1	VV_2	VV_3	VV_4	VV_5	VV_6
	0	+1	V_{31}	V_0	V_{31}	V_0	V_{31}	V_0	V_{31}	V_0	V_{31}	V_0
		−1	V_{31}	V_0	V_{31}	V_0	V_{31}	V_0	V_{31}	V_0	V_{31}	V_0

4. Experimental Work

The performance of the DTC controller using VVs was experimentally tested in healthy and open-phase operation, with no reconfiguration of the controller. The experimental test bench is shown in Figure 4. It was composed of a five-phase IM fed by two conventional three-phase VSCs from Semikron. The DC-link was set to 300 V by an external DC power supply. The controller was programmed on a MSK28335 board and a TMS320F28335 microcontroller. The mechanical speed was measured by a digital encoder and the microcontroller's peripherals. Additionally, a variable load torque was applied by a mechanically coupled DC machine. An open-phase fault condition in phase 'a' was emulated in the provided tests by opening a power relay connected in series with the phase. The IM characteristics are listed in Table 2; the reference stator flux was set to 0.389 Wb, the sampling time used was 100 µs and the hysteresis bands of the torque and flux regulators were fixed at 1.06% and

1.29% of the rated values, respectively. Note that a limitation in the integral part of the controller and an anti-windup scheme were included in the implemented control algorithm to prevent integration windup in the PI-based speed controller when the actuator is saturated.

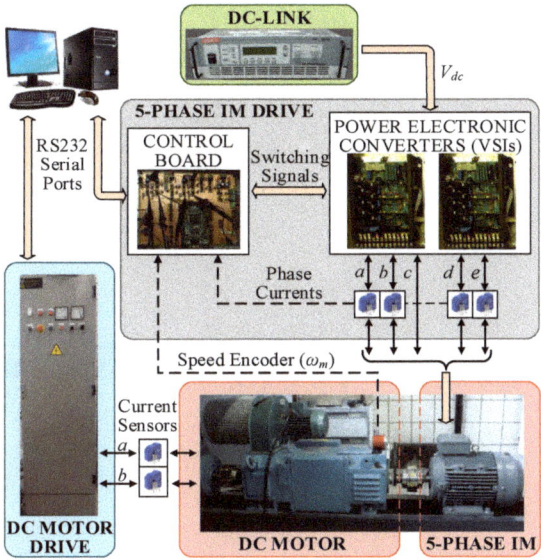

Figure 4. Experimental system.

Table 2. Electrical and mechanical parameters of the five-phase IM.

Parameter	Value	Unit
Stator resistance, R_s	12.85	Ω
Rotor resistance, R_r	4.80	Ω
Stator leakage inductance, L_{ls}	79.93	mH
Rotor leakage inductance, L_{lr}	79.93	mH
Mutual inductance, L_m	681.7	mH
Moment of inertia, J	0.02	kg·m²
Number of pole pairs, p	3	-
Rated torque, T_n	4.70	N·m
Rated stator flux, λ_s^*	0.389	Wb

The transition from prefault to postfault operation was first analyzed when the speed was maintained at 500 rpm and a load torque around 60% of the nominal one was imposed. The results obtained are shown in Figures 5 and 6. Although the speed decreases a little when the fault occurs, it can be observed that the DTC controller is capable of controlling the speed and the torque of the system even using the prefault voltage vectors and system model (see Figure 5a,b). Since the MMF remains the same in both healthy and faulty operation, a circular trajectory is obtained in the α–β currents, as can be seen in Figure 5c. However, a horizontal line appears in the x–y plane because $i_{xs} = -i_{\alpha s}$ and $i_{ys} = 0$. In accordance with the well-known minimum copper loss criterion [23], stator phase currents 'b' and 'e' are equal in magnitude but present unequal peak values compared to 'c' and 'd' phase currents, as shown in Figure 6.

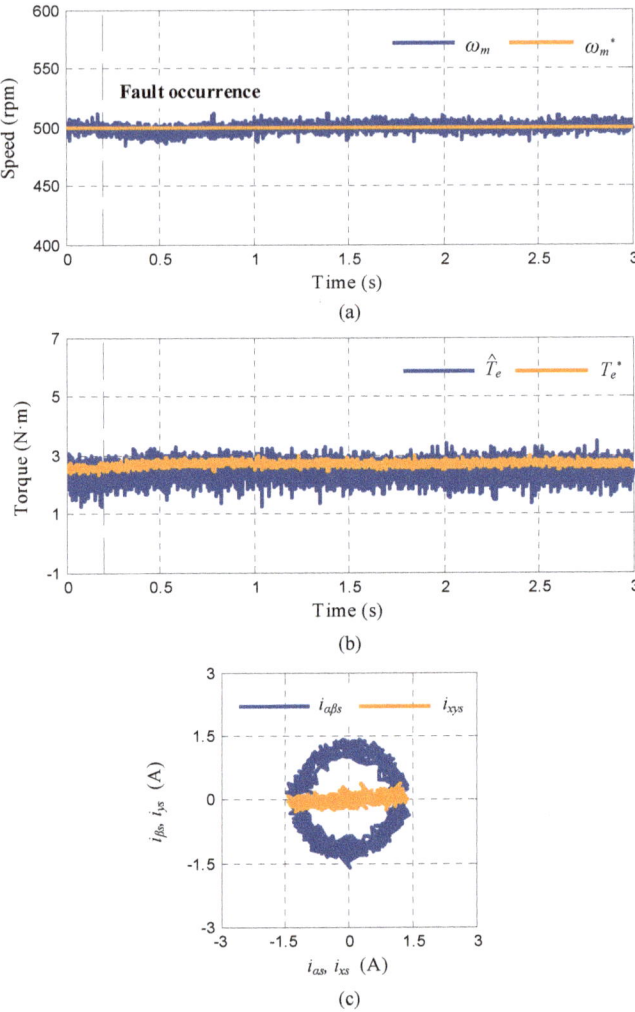

Figure 5. Transition from prefault to postfault operation using a DTC controller with VVs. The fault occurs at $t = 0.2$ s; the speed is set to 500 rpm and a load torque around 60% of the nominal one is applied. (**a**) Measured speed (ω_m) and its reference (ω_m^*); (**b**) estimated torque ($\hat{T}_e$) and its reference (T_e^*); and (**c**) stator currents in the α–β and x–y planes ($i_{\alpha\beta s}$, i_{xys}).

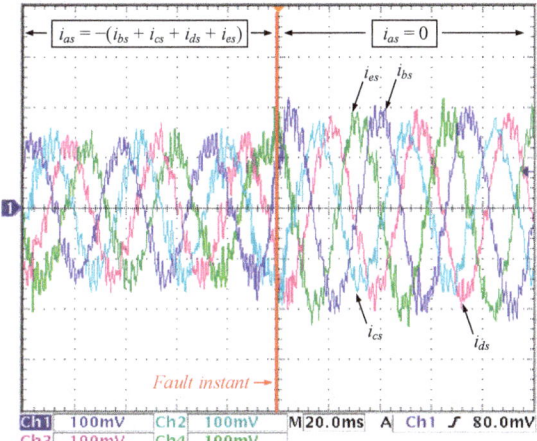

Figure 6. Phase currents i_{bs}, i_{cs}, i_{ds} and i_{es} in the transition from prefault to postfault operation using DTC with VVs (the faulty phase, i_{as}, has similar behavior before the fault appears, being null after the fault). The fault occurs at $t = 0.2$ s, the speed is set to 500 rpm and a load torque around 60% of the nominal one is applied.

Next, the dynamic performance of the DTC controller in faulty operation was analyzed. A speed step from 0 to 500 rpm was forced with a null-load torque condition. Figure 7 shows the speed, torque and flux performance (Figure 7a–c), and it can be seen that the DTC scheme provides an accurate tracking performance of all variables. It is necessary to highlight that the electrical torque is limited to 3 N·m, since nominal torque is not reachable in a postfault situation. The phase current evolution is also depicted in Figure 7d.

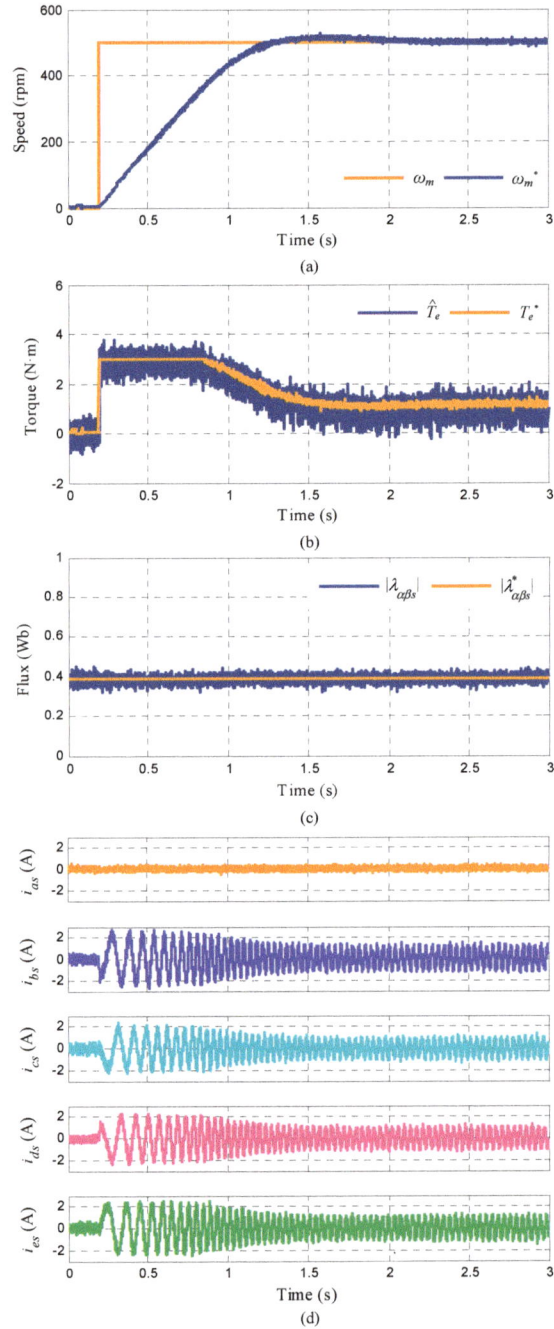

Figure 7. Speed step response in open-phase fault operation of the proposed DTC controller. The reference speed is changed from 0 to 500 rpm at $t = 0.2$ s and no load torque is imposed. (**a**) Measured speed (ω_m) and its reference (ω_m^*); (**b**) estimated torque ($\hat{T}_e$) and its reference (T_e^*); (**c**) stator flux in the α–β plane; (**d**) phase currents.

The rejection properties of the system are also studied and the results presented in Figure 8. A change in the load torque is applied at $t = 0.8$ s while the speed reference is fixed to 500 rpm. The tracking performance of the electrical torque is quite accurate, even in the transitory. On the other hand, a decrease in the rotor speed is observed when the fault occurs, but the controller successfully manages the disturbance in a short period of time.

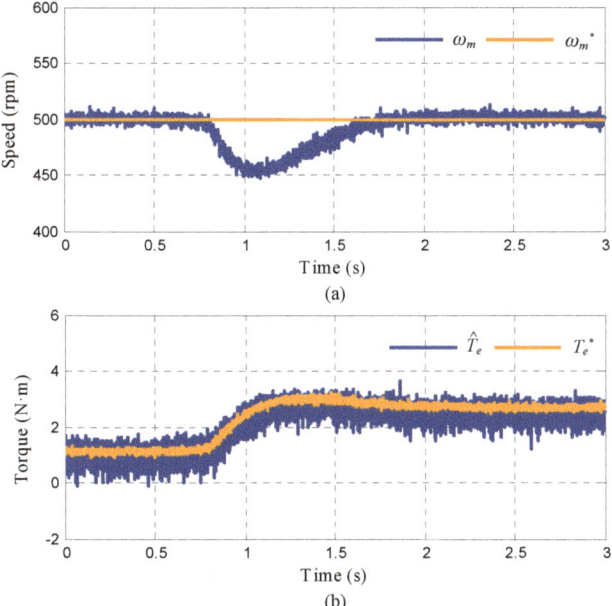

Figure 8. Load-torque rejection response in open-phase fault operation of the DTC controller with VVs. A change from 0 N·m to around 60% of the nominal torque is applied at $t = 0.8$ s. (**a**) The measured speed (ω_m) and its reference (ω_m^*). (**b**) The estimated torque ($\hat{T}_e$) and its reference (T_e^*).

Finally, a reversal test is reproduced and shown in Figure 9. In this experiment, the reference speed is first settled in at 500 rpm and then changed to −500 rpm at $t = 0.2$ s. The system is again operated in open-phase fault condition, with a null load torque applied to the system. An appropriate tracking of the speed, the estimated torque and the stator flux is obtained, with a good crossing by zero performance. The evolution of phase currents during this test is depicted in Figure 9d.

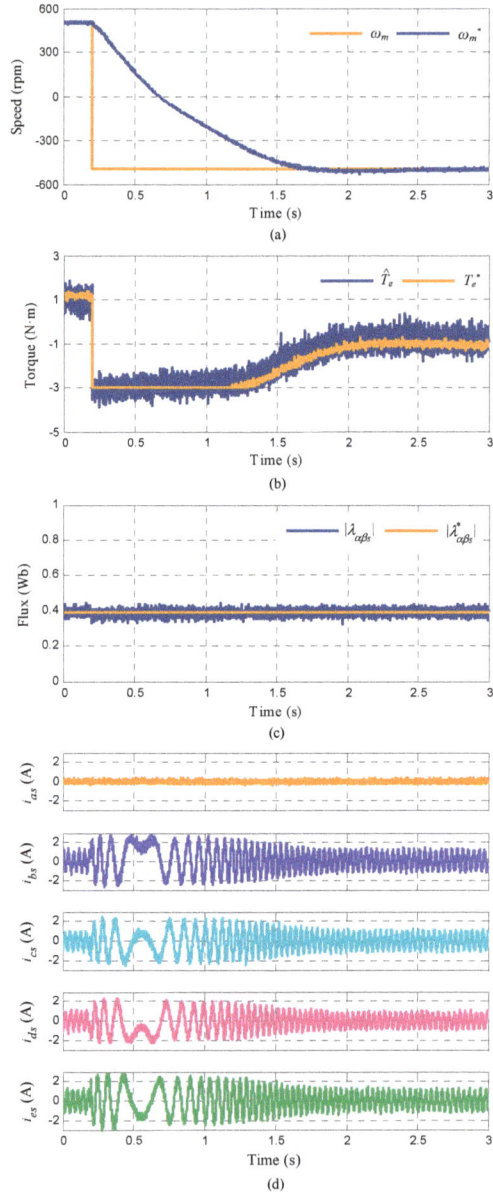

Figure 9. Speed reversal test in open-phase fault operation of the proposed DTC controller. The reference speed is changed from 500 to −500 rpm at $t = 0.2$ s and no load torque is imposed. (**a**) Measured and reference speed; (**b**) estimated torque ($\hat{T}_e$) and its reference (T_e^*); (**c**) stator flux in the α–β plane; (**d**) phase currents.

5. Conclusions

The recent interest in multiphase drives for particular industry applications is mainly based on their inherent fault-tolerant characteristics. However, the complexity of the applied controllers, which usually require extensive reconfiguration to manage the faulty operation, hinders this interest. This work shows that VVs combined with direct controllers can overcome this difficulty and provide a natural fault-tolerant capability for multiphase drives. Indeed, the idea, previously introduced for FCS-MPC controllers, is here experimentally validated when DTC is applied, where VVs allow the use of the same voltage vectors, lookup tables and control scheme in pre- and postfault situations.

Author Contributions: Conceptualization and methodology, F.B. and M.J.D.; software, M.B., P.S. and I.G.-P.; validation, all authors; formal analysis and investigation, all authors; resources, all authors; data curation, all authors; writing—original draft preparation, F.B.; writing—review and editing, all authors; visualization, all authors; supervision, M.J.D. and F.B.; project administration, F.B.; funding acquisition, I.G.-P. and F.B.

Funding: This research was funded by the Spanish Government.

Conflicts of Interest: The authors declare no conflict of interest.

References

1. Levi, E.; Barrero, F.; Duran, M.J. Multiphase Machines and Drives-Revisited. *IEEE Trans. Ind. Electron.* **2016**, *63*, 429–432. [CrossRef]
2. Duran, M.J.; Levi, E.; Barrero, F. Multiphase Electric Drives: Introduction. In *Wiley Encyclopedia of Electrical and Electronics Engineering*; John Wiley & Sons: Hoboken, NJ, USA, 2017.
3. Mohammadpour, A.; Parsa, L. A Unified Fault-Tolerant Current Control Approach for Five-Phase PM Motors with Trapezoidal Back EMF under Different Stator Winding Connections. *IEEE Trans. Power Electron.* **2013**, *28*, 3517–3527. [CrossRef]
4. Mohammadpour, A.; Sadeghi, S.; Parsa, L. A Generalized Fault-Tolerant Control Strategy for Five-Phase PM Motor Drives Considering Star, Pentagon, and Pentacle Connections of Stator Windings. *IEEE Trans. Ind. Electron.* **2014**, *61*, 63–75. [CrossRef]
5. Duran, M.J.; Barrero, F. Recent Advances in the Design, Modeling and Control of Multiphase Machines-Part 2. *IEEE Trans. Ind. Electron.* **2016**, *63*, 459–468. [CrossRef]
6. Sadeghi, S.; Guo, L.; Toliyat, H.A.; Parsa, L. Wide Operational Speed Range of Five-Phase Permanent Magnet Machines by Using Different Stator Winding Configurations. *IEEE Trans. Ind. Electron.* **2012**, *59*, 2621–2631. [CrossRef]
7. Mercorelli, P.; Kubasiak, N.; Liu, S. Multilevel Bridge Governor by using Model Predictive Control in Wavelet Packets for Tracking Trajectories. In Proceedings of the IEEE International Conference on Robotics and Automation, New Orleans, LA, USA, 26 April–1 May 2004; Volume 4, pp. 4079–4084.
8. Barrero, F.; Duran, M.J. Recent Advances in the Design, Modeling and Control of Multiphase Machines-Part 1. *IEEE Trans. Ind. Electron.* **2016**, *63*, 449–458. [CrossRef]
9. Barcaro, M.; Bianchi, N.; Magnussen, F. Faulty Operations of a PM Fractional-Slot Machine with a Dual Three-Phase Winding. *IEEE Trans. Ind. Electron.* **2011**, *58*, 3825–3832. [CrossRef]
10. Nguyen, N.K.; Meinguet, F.; Semail, E.; Kestelyn, X. Fault-Tolerant Operation of an Open-End Winding Five-Phase PMSM Drive with Short-Circuit Inverter Fault. *IEEE Trans. Ind. Electron.* **2016**, *63*, 595–605. [CrossRef]
11. Zarri, L.; Mengoni, M.; Gritli, Y.; Tani, A.; Filippetti, F.; Serra, G.; Casadei, D. Detection and Localization of Stator Resistance Dissymmetry Based on Multiple Reference Frame Controllers in Multiphase Induction Motor Drives. *IEEE Trans. Ind. Electron.* **2013**, *60*, 3506–3518. [CrossRef]
12. Abdel-Khalik, A.S.; Masoud, M.I.; Ahmed, S.; Massoud, A. Calculation of Derating Factors based on Steady-State Unbalanced Multiphase Induction Machine Model under Open Phase(s) and Optimal Winding Currents. *Electr. Power Syst. Res.* **2014**, *106*, 214–225. [CrossRef]
13. Guzman, H.; Barrero, F.; Duran, M.J. IGBT-Gating Failure Effect on a Fault-Tolerant Predictive Current-Controlled Five-Phase Induction Motor Drive. *IEEE Trans. Ind. Electron.* **2015**, *62*, 15–20. [CrossRef]
14. Dwari, S.; Parsa, L. An Optimal Control Technique for Multiphase PM Machines under Open-Circuit Faults. *IEEE Trans. Ind. Electron.* **2008**, *55*, 1988–1995. [CrossRef]
15. Gonzalez-Prieto, I.; Duran, M.J.; Aciego, J.J.; Martin, C.; Barrero, F. Model Predictive Control of Six-Phase Induction Motor Drives using Virtual Voltage Vectors. *IEEE Trans. Ind. Electron.* **2018**, *65*, 27–37. [CrossRef]

16. Gonzalez-Prieto, I.; Duran, M.J.; Bermúdez, M.; Barrero, F.; Martín, C. Assessment of Virtual-Voltage-based Model Predictive Controllers in Six-phase Drives under Open-Phase Faults. *IEEE J. Emerg. Sel. Top. Power Electron.* **2019**, 1. [CrossRef]
17. Jahns, T.M. Improved Reliability in Solid-State AC Drives by means of Multiple Independent Phase Drive Units. *IEEE Trans. Ind. Appl.* **1980**, *16*, 321–331. [CrossRef]
18. Yousefi-Talouki, A.; Gholamian, S.A.; Yousefi-Talouki, M.; Ilka, R.; Radan, A. Harmonic Elimination in Switching Table-based Direct Torque Control of Five-Phase PMSM using Matrix Converter. *IEEE Symp. Humanit. Sci. Eng. Res. Kuala Lumpur* **2012**, 777–782.
19. Zheng, L.; Fletcher, J.E.; Williams, B.W.; He, X. A Novel Direct Torque Control Scheme for a Sensorless Five-Phase Induction Motor Drive. *IEEE Trans. Ind. Electron.* **2011**, *58*, 503–513. [CrossRef]
20. Gao, L.; Fletcher, J.E.; Zheng, L. Low speed control improvements for a 2-level 5-phase inverter-fed induction machine using classic direct torque control. *IEEE Trans. Ind. Electron.* **2011**, *58*, 2744–2754. [CrossRef]
21. Bermudez, M.; Gonzalez-Prieto, I.; Barrero, F.; Guzman, H.; Duran, M.J.; Kestelyn, X. Open-phase fault-tolerant direct torque control technique for five-phase induction motor drives. *IEEE Trans. Ind. Electron.* **2016**, *64*, 902–911. [CrossRef]
22. Riveros, J.A.; Barrero, F.; Levi, E.; Duran, M.; Toral, S.; Jones, M. Variable-speed five-phase induction motor drive based on predictive torque control. *IEEE Trans. Ind. Electron.* **2013**, *60*, 2957–2968. [CrossRef]
23. Fu, J.R.; Lipo, T.A. Disturbance-free operation of a multiphase current-regulated motor drive with an opened phase. *IEEE Trans. Ind. Appl.* **1994**, *30*, 1267–1274.

© 2019 by the authors. Licensee MDPI, Basel, Switzerland. This article is an open access article distributed under the terms and conditions of the Creative Commons Attribution (CC BY) license (http://creativecommons.org/licenses/by/4.0/).

Article

Prediction of PWM-Induced Current Ripple in Subdivided Stator Windings Using Admittance Analysis

Antoine Cizeron [1,2,*], Javier Ojeda [2], Eric Labouré [1] and Olivier Béthoux [1]

[1] GeePs | Group of Electrical Engineering—Paris, CNRS, CentraleSupélec, University Paris-Sud, Université Paris-Saclay, Sorbonne Université, 3 & 11 rue Joliot-Curie, Plateau de Moulon 91192 Gif-sur-Yvette CEDEX, France; eric.laboure@geeps.centralesupelec.fr (E.L.); olivier.bethoux@geeps.centralesupelec.fr (O.B.)

[2] SATIE | Systèmes et Applications des Technologies de l'Information et de l'Energie, ENS Paris-Saclay, CNRS, Université Paris-Saclay, 94235 Cachan, France; javier.ojeda@satie.ens-cachan.fr

* Correspondence: antoine.cizeron@geeps.centralesupelec.fr

Received: 30 September 2019; Accepted: 28 October 2019; Published: 21 November 2019

Abstract: Subdividing stator winding is a way to lower the DC link voltage value in electric drives and reduce the stress on motor insulation. Coupled windings sharing the same stator teeth are modelled in order to evaluate the link between voltages disparities and current ripple. This paper provides an assessment of current ripple rise in the subdivided windings compared to ordinary topologies through the use of a basic inductive model. A method for PWM-Induced current ripple and high-frequency loss estimation based on admittance measurements is developed and experimentally validated. The use of this subdivided structure does not induce more than a 10% rise of the PWM-induced current ripple compared to a standard winding structure.

Keywords: electric drives; winding configuration; modelling; pulse width modulation; current ripple; high-frequency losses

1. Introduction

Sizing of electrical powertrains in transportation applications must deal with severe size and mass constraints [1]. Increasing the switching frequency appears to be a solution to improve power integration as it permits smaller passive components [2]. The high slew rate of GaN or SiC-based switches enables the high switching frequencies required although a trade-off between low switching losses and high electromagnetic interferences must be considered [3,4]. In electric drive applications, high dv/dt rates may cause insulation degradation [5,6] and bearing wear [7] leading to a shorter lifetime. Traction chains in electric vehicles (EV) are supplied by a DC bus usually operating in a voltage range of between 300 V and 600 V [1]. DC-link voltage reduction can be investigated to reduce insulation stress and the cost of DC-bus capacitors [8]. Reconfiguration of windings can extend the rated power or the speed range for a given DC-link voltage value [9,10]. Such a change in motor windings may also improve fault tolerance [11,12].Using a drive DC voltage of under 60 V is recognised by international safety certifications (such as CE mark) as reducing the potential danger to the equipment user. It also allows the use of topologies as proposed by [13], to take advantage of highly parallel configurations of battery cells. The approach of this paper is to explore a new winding configuration that allows electric drive motors to operate at low DC voltage without impacting their electromagnetic design. This concept is related to a recent patent [14]. Based on this, the powertrain supply subdivision enables use of a low DC bus voltage and improve the resiliency of the system as a whole. Multiphase drives are also a way to improve the rated power or the torque quality [15,16] for given conductors and phase voltage. Fault tolerance is also an advantage of multiphasing [17]. Several multiphase topologies are detailed in [18]. The proposed topology extends multiphasing concepts beyond classical winding reconfiguration and draws on fractionation granularity at turn level. To start with, this paper examines electrical consequences of a stator coil subdivision by two and proposes a method to estimate current ripple and power losses in such highly coupled systems.

The proposed topology is based on stator windings subdivision. The basic principle is to deconstruct the initial motor windings in n subdivided windings supplied by n individual and independent inverters instead of one; hence, the machine coil as well as the copper design remains unchanged. However the global electric drive is highly modified, leading to low voltage supply and a high degree of freedom. A high voltage designed machine is taken as an example. Each machine armature winding is divided by n and all subdivided windings are independently controlled so that one phase is broken down in n modules. As an example, one phase of a three-phase machine is subdivided by three as presented in Figure 1. Winding A1 visible in Figure 1a is divided as shown in Figure 1b. The resulting windings A11, A12 and A13 are wound around the same stator tooth symbolised by dotted lines in Figure 2. Only the supply of the subdivided windings is modified compared to the standard machine. The motor magnetic core remains unchanged from its initial design. The power supplies of the subdivided windings are parallelized and fed by the same DC bus. The DC-link voltage is hence divided by three compared to the ordinary case (Figure 2a). Thus, a high voltage designed machine normally fed by a high voltage inverter is turned into a low voltage drive combining a low voltage highly subdivided machine and a fractioned inverter. The motor insulation stress is therefore reduced by n and also the dv/dt switching slew rate is much even distributed along the subdivided core avoiding a classic voltage over stress at the coil end turns [19]. The benefit is obviously a significant increase in lifetime since insulating ageing is a key factor in motor failures [20]. In addition, the derived drive has new degrees of freedom. Nevertheless, many of them have to be strictly managed in order to precisely control the various currents in the sub-coils located in the same armature slot. These currents are closely linked by a strong magnetic coupling. To avoid any significant current ripple, it is mandatory to ascertain to what extent the sub-coils applied voltages may differ. Taking into account the fact that the self-inductance of each subdivided winding is reduced, the present work carefully assesses the current ripple rise induced by the inevitable voltages discrepancies. The aim of this paper is to investigate the range value of the new architecture degrees of freedom in order to determine whether this range is relevant to the actual technological capability. This addresses the key feasibility issue of implementing this concept.

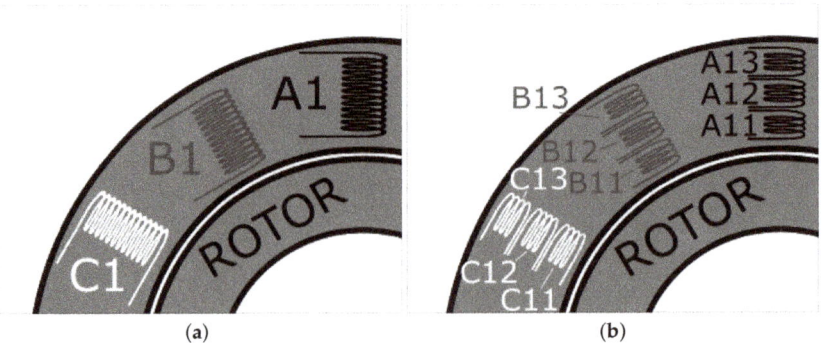

Figure 1. 3-phase, 4 teeth per phase machine fractionation by $n = 3$: (**a**) Ordinary case, (**b**) Subdivided case.

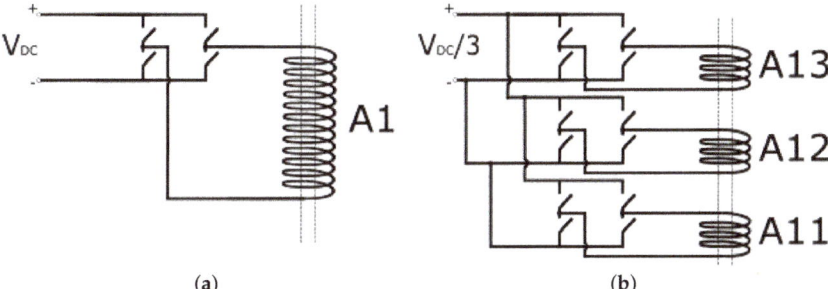

Figure 2. Phase A supply system subdivision by $n = 3$. (**a**) Ordinary case, (**b**) Subdivided case.

This paper focuses on a single tooth of the first phase. The initial coil wound around this tooth is divided into two individual sub-coils (Figure 3). Each subdivided coil obtained is supplied by its own inverter (Figure 2b). The study focuses on one switching period, which is the relevant temporal scale for this highly magnetically coupled coils. As winding subdivision is not common, models have to be built in order to evaluate the current ripple. Finding analytical solutions in study of Pulse Width Modulation (PWM) effects on current ripple is interesting as a finite elements analysis model would require large computing time [21]. Conversely, designing an analytical formulation requires low computing effort and permits rapid calculation and the ability to derive the main whys and wherefores. This latter approach is therefore used in this study. However, laminated steel behaviour at high switching frequencies is not completely understood, particularly in this context where the various coils voltages stimulate the leakage inductance of sub-windings at frequencies where magnetic field is no longer penetrating materials. A new method is therefore proposed in order to estimate current ripples in windings under PWM voltage stimuli.

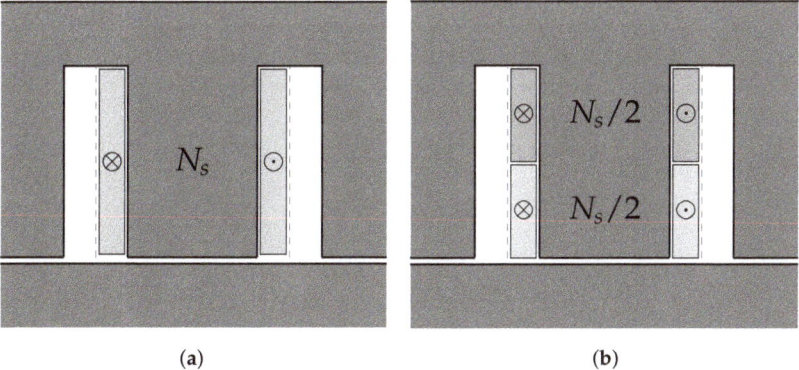

Figure 3. Subdivision by $n = 2$ of a winding with N_s number of turns around a stator tooth (**a**) Ordinary case, (**b**) Subdivided case.

This work aims to provide a consistent model in terms of current ripple evaluation and high-frequency additional power loss assessment in the specific context of interactions between sub-windings operating in the new fractionated motor drive. The paper structure is as follows: Section 1 introduces the scientific and technological contexts leading to winding subdivision and considers them in a wider approach of multiphase drive and winding reconfiguration. This introductory section highlights the critical importance of addressing current ripple estimation and power losses assessment of the original structure under study. Section 2 uses a first basic inductive model to investigate the effect of the differential mode between voltage applied to the sub-windings. To enhance the first model, Section 3 takes a frequency approach based on a wide band admittance measurement. It enables to get a more specific analysis of current ripples and high-frequency power losses under real PWM voltages. In Section 4, these theoretical developments are tested using an experimental setup made of two independent inverters supplying two subdivided windings located in the same magnetic core. The experimental protocol is fully described and the related results are commented; they validate the trends identified by the analytical study. In section 5 the findings of this comprehensive study are placed in the proper perspective of the studied motor drive architecture. It highlights the scope of the present findings and shows all the important parameters required to evolve from a proof of concept to a first prototype. Finally, conclusions summarising key points are presented in Section 6 and complemented by perspectives on future work.

2. Current Ripple Assessment

2.1. Defaults in Winding Subdivision Use

Two windings from the same phase located in one common stator tooth are considered. They result in subdividing a winding of an ordinary electrical machine in two windings with the same number of turns (Figure 3). The airgap influence on electrical disparities between both windings is not considered in this initial study. The aim of this section is to assess the impact of the winding subdivision concept on the winding current ripple using standard PWM voltages. As the studied innovative concept consists in splitting a standard winding into several sub-windings, the adopted performance criterion is named the Current Ripple Ratio CRR, and defined as:

$$CRR = \frac{\Delta i}{\Delta i_0} \qquad (1)$$

where Δi is the current ripple in the subdivided case (Figure 4b) and Δi_0 is the current ripple in a classical architecture (Figure 4a). In order to compare ordinary and subdivided topologies, the magnetic core and the fundamental PWM frequency $F_s = 1/T_s$ are fixed. The comparison is made during a single switching period. The winding are supposed to be in a no-load motor configuration as this is the worst case in terms of current ripple, as the magnitude of voltages applied to windings are not decreased by the electromotive force. The magnetic circuit polarisation due to the low-frequency current component is not taken into account because its impact on current ripple is similar in both topologies. The purpose of the study is precisely to understand how the new windings supply impacts the current waveforms and to assess the relative shapes.

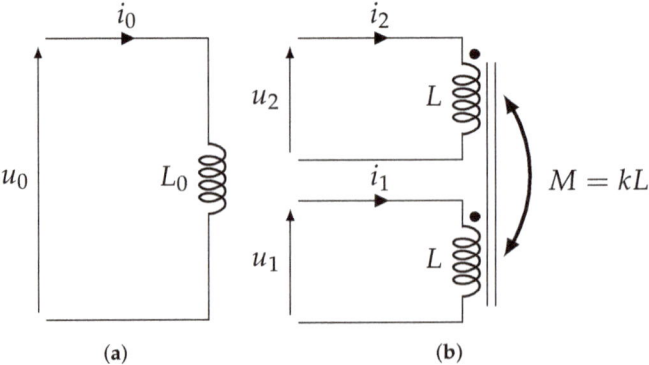

Figure 4. Link between two inductive models (**a**) Ordinary case, (**b**) Subdivided case.

To compute an analytical expression of the Current Ripple Ratio CRR, this section considers a basic purely inductive model of the subdivided windings. This kind of inductive model is widely used in the study of interleaved converters [22]. In the present case, the mutual parameter is positive unlike in the multicell converter one. In order to separate the effects related to disparities between the two sub-coils voltages from those related to electrical discrepancies between both subdivided windings, a symmetrical model is considered ($L \approx L_1 \approx L_2$). Coupled circuit described in Figure 4b leads to :

$$\begin{pmatrix} u_1 \\ u_2 \end{pmatrix} = \begin{pmatrix} L & kL \\ kL & L \end{pmatrix} \cdot \frac{d}{dt} \begin{pmatrix} i_1 \\ i_2 \end{pmatrix} \quad (2)$$

where k is the coupling factor between two windings. This system relates voltages to current slope as:

$$\frac{d}{dt} \begin{pmatrix} i_1 \\ i_2 \end{pmatrix} = \frac{1}{(1-k^2)L} \cdot \begin{pmatrix} 1 & -k \\ -k & 1 \end{pmatrix} \cdot \begin{pmatrix} u_1 \\ u_2 \end{pmatrix} \quad (3)$$

This model suggests the relationship between a classical winding configuration (Figure 4a) and a fractioned combination (Figure 4b) excited by two identical voltages ($u_1 = u_2$):

$$u_0 = L_0 \cdot \frac{di_0}{dt} \quad \text{and} \quad (u_1 + u_2) = L(1+k) \cdot \frac{d(i_1 + i_2)}{dt} \quad (4)$$

From power electronics point of view, $u_0 = (u_1 + u_2)$ because a subdivided winding requires half initial voltage (Figure 2) and from machiner point of view, Ampere-turns are kept constant using $i_1 = i_2 = i_0$ as total number of coil turns remains constant (Figure 3). Applied to (4), this leads to

$$L_0 = 2 \cdot L \cdot (1+k) \quad (5)$$

This relation is used to evaluate CRR due to differences between the sub-windings voltages compared to ordinary winding case. Models are equivalent when $|u_1| = |u_2| = V_{DC}$ and $|u_0| = 2 \cdot V_{DC}$. Resulting current shapes are shown in Figure 5a. V_{DC} is the DC-link voltage associated to inverters supplying sub-windings. DC-link voltage associated to an ordinary winding is twice this value as shown in Figure 2. In identical voltage case, (Figure 5a), Δi current ripple in each subdivided winding is equal to Δi_0, which is:

$$\Delta i_0 = \frac{2V_{DC}}{L_0} \cdot \frac{T_s}{2} = \frac{V_{DC} T_s}{2 \cdot L(1+k)} \quad (6)$$

Considering that both sub-windings are supplied by their own independent inverters as shown in Figure 2b, u_1 and u_2 may present a time delay or a duty-cycle difference in normal operation mode. Indeed, the propagation time in the switch drivers may slightly differ and similarly the sub-windings discrepancies may induce a little duty-cycle difference to substantially equalise both average currents. These two different aspects are investigated through a basic inductive model represented in Figure 4b. In both cases, namely time delay and duty-cycle difference, the current ripple is computed in order to evaluate CRR induced by the independent control of the sub-windings.

2.2. Delay

This part details the CRR expression while the single voltages face a relative delay which, by definition, does not occur in the standard case (Figure 4a). Two centred PWM characterised by a similar duty cycle are considered: the study is limited to the worst case consisting in $\alpha_1 = \alpha_2 = 0.5$. Considering the fact that the propagation time in each driver may present disparities, a time delay τ between the sub-windings voltages may appear as depicted in Figure 5b. When such a delay occurs, four time domains $\{D_1, D_2, D_3, D_4\}$ may be distinguished during one switching period. For each domain, the related current slope is computed with (3) and their values are summarised in Table 1.

Table 1. Voltage and current slope under a delay.

	D_1	D_2	D_3	D_4
u_1	$+V_{DC}$	$+V_{DC}$	$-V_{DC}$	$-V_{DC}$
u_2	$-V_{DC}$	$+V_{DC}$	$+V_{DC}$	$-V_{DC}$
$\dfrac{di_1}{dt}$	$\dfrac{V_{DC}}{L(1-k)}$	$\dfrac{V_{DC}}{L(1+k)}$	$\dfrac{-V_{DC}}{L(1-k)}$	$\dfrac{-V_{DC}}{L(1+k)}$
$\dfrac{di_2}{dt}$	$\dfrac{-V_{DC}}{L(1-k)}$	$\dfrac{V_{DC}}{L(1+k)}$	$\dfrac{V_{DC}}{L(1-k)}$	$\dfrac{-V_{DC}}{L(1+k)}$

Following these current evolutions, i_1 and i_2 shapes are shown in Figure 5b. Obviousuly both windings face an additional current ripple when a time delay occurs. Differential mode $u_1 = -u_2$ induced during D_1 and D_3 leads to a high current slope in both windings. Consequently, current shapes are modified compared to the standard synchronised case (Figure 5a); i_2 is in phase opposition with u_2 while i_1 and u_1 are in phase accordance. It demonstrates a new and adverse power flow from the first to the second winding. This power flow between sub-windings is an unwanted side effect in a motor context as it only causes additional losses in the conductors. In this way, this suggests that the control system has to synchronise the sub-windings voltages in order to avoid any differential mode induced by a time delay. Nevertheless, a residual slight delay may appear which clearly relies on technological aspects. To assess the minimum–maximum acceptable delay range, CRR is computed regarding $\left(\dfrac{|\tau|}{T_s}\right)$ dimensionless ratio. Deriving from Figure 5b and Table 1, the current ripple due to a voltage delay is:

$$\Delta i = \frac{V_{DC} \cdot ((1-k)T_s + 4k|\tau|)}{2L(1-k^2)} \tag{7}$$

$$\Delta i = V_{DC} \cdot \left[\frac{T_s}{2L(1+k)} + \frac{4k|\tau|}{2L(1-k^2)}\right] \tag{8}$$

Obviously, the minimum value of Δi corresponds to the standard case Δi_0 (6). Therefore, it derives:

$$\Delta i = \Delta i_0 + \frac{V_{DC}}{2L(1+k)} \cdot \frac{4k|\tau|}{(1-k)} \tag{9}$$

Finally, CRR due to the time delay between subdivided voltages is expressed as:

$$CRR = \left(\frac{\Delta i}{\Delta i_0}\right) = 1 + \frac{4k}{(1-k)} \cdot \left(\frac{|\tau|}{T_s}\right) \tag{10}$$

Equation (10) establishes a mathematical relationship between relative delay and current ripple rise based on a single parameter, namely k the coupling factor between both sub-coils. Equation (10) shows that, in switching frequency range, having a low coupling factor lowers current ripple rise. This limits the impact of the voltages delay on CRR. Conversely, at low range frequencies the sub-coils leakage inductances are expected to be small in order to produce a large magnetic field in the air-gap leading to a coupling factor close to 1 (at this frequency range, i.e., from 0 Hz to hundreds of Hz).

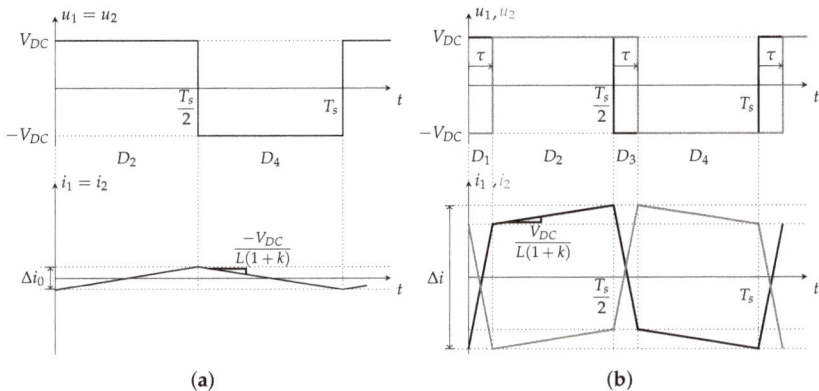

Figure 5. Voltage and Current Evolution under (**a**) synchronised control signals, (**b**) a delay between control signals.

Section 4 shows that the coupling factor of two adjacent sub-coils is roughly 0.9 at frequencies around 25 kHz in the studied proof of concept instance. Supposing winding subdivision concept is regarded as acceptable for a current ripple rise lower than 10% (i.e., $CRR = 1.1$), sub-coils voltages should not present a delay higher than $\tau = 110$ ns. It is highly reasonable to consider that power switches drivers propagation time discrepancies are lower than 50 ns. Hence, present technology clearly guarantees a time delay of less than 110 ns. Therefore, residual technological delay appearing between voltages is not a problem in the validation of winding subdivision concept.

2.3. Duty-Cycle Difference

As current slaving in both coils may be different because of their electrical parameters disparities, it is now assumed that the duty-cycles applied to each converter may differ. Similarly to delay study, duty-cycles difference doesn't occur in the standard case (Figure 4a). Purely inductive model also enables to investigate the impacts of duty-cycles difference on CRR (1) while supposing $\tau = 0$. Sub-coils voltages are depicted in Figure 6. In this case, four time domains $\{D_1, D_2, D_3, D_4\}$ appear during one switching period. For each domain, the related current slope is computed using (3) and shown in Table 2.

Table 2. Voltage and current slope under a duty-cycle difference.

	D_1	D_2	D_3	D_4
u_1	$-V_{DC}$	$-V_{DC}$	$+V_{DC}$	$-V_{DC}$
u_2	$-V_{DC}$	$+V_{DC}$	$+V_{DC}$	$+V_{DC}$
$\frac{di_1}{dt}$	$\frac{-V_{DC}}{L(1+k)}$	$\frac{-V_{DC}}{L(1-k)}$	$\frac{V_{DC}}{L(1+k)}$	$\frac{-V_{DC}}{L(1-k)}$
$\frac{di_2}{dt}$	$\frac{-V_{DC}}{L(1+k)}$	$\frac{V_{DC}}{L(1-k)}$	$\frac{V_{DC}}{L(1+k)}$	$\frac{V_{DC}}{L(1-k)}$

This leads to i_1 and i_2 waveforms represented in Figure 6. Based on the purely inductive model, voltages may present non-zero average values causing sub-coils currents to diverge. As this section focuses on current ripple, low frequency current evolutions (11) are suppressed to exclusively capture high-frequency current component $\tilde{i}_1$ and $\tilde{i}_2$. Technically, low frequency current component converges to a permanent value of current related to windings resistance and high-frequency component corresponds to steady state behaviour. This is experimentally verified in Section 2.3.

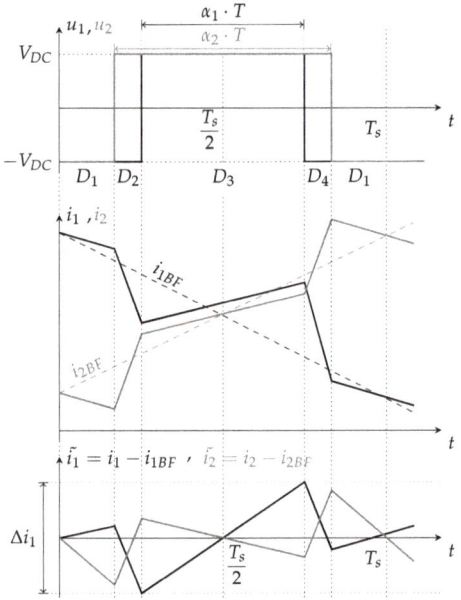

Figure 6. Voltage and Current evolution under a duty-cycle difference $\alpha_1 < \alpha_2$.

Using (3) and knowing that average voltages are $u_{1BF} = V_{DC} \cdot (2\alpha_1 - 1)$ and $u_{2BF} = V_{DC} \cdot (2\alpha_2 - 1)$, low frequency currents waveforms can be derived as:

$$\begin{cases} \dfrac{di_{1BF}}{dt} = \dfrac{V_{DC}}{(1-k^2)L} \cdot [2(\alpha_1 - k\alpha_2) - (1-k)] \\ \\ \dfrac{di_{2BF}}{dt} = \dfrac{V_{DC}}{(1-k^2)L} \cdot [2(\alpha_2 - k\alpha_1) - (1-k)] \end{cases} \quad (11)$$

Subtracting the low frequency component (11) to the global current estimated from inductive model (Table 2) leads to extract high-frequency current ripple during each time domain $\{D_1, D_2, D_3, D_4\}$ (12). As both windings are considered symmetrical, the current ripple is simply computed in the first winding. In $\alpha_1 < \alpha_2$ case specific phases D_1 and D_3 last $(1-\alpha_2)T_s$ and $\alpha_1 T_s$, respectively. The related current ripple can be written as:

$$\begin{cases} (\Delta i_1)_{D_1} = \left| \left(\dfrac{di_1}{dt} - \dfrac{di_{1BF}}{dt} \right)_{D_1} \right| \cdot (1-\alpha_2)T_s = \dfrac{2V_{DC}}{L(1+k)} \cdot \left| \dfrac{\alpha_1 - k\alpha_2}{1-k} \right| \cdot (1-\alpha_2)T_s \\ \\ (\Delta i_1)_{D_3} = \left| \left(\dfrac{di_1}{dt} - \dfrac{di_{1BF}}{dt} \right)_{D_3} \right| \cdot \alpha_1 T_s = \dfrac{2V_{DC}}{L(1+k)} \cdot \left| 1 - \dfrac{\alpha_1 - k\alpha_2}{1-k} \right| \cdot \alpha_1 T_s \end{cases} \quad (12)$$

with $f_k(\alpha_1, \alpha_2) = \dfrac{\alpha_1 - k\alpha_2}{1-k}$ et $\Delta i_0 = \dfrac{V_{DC} T_s}{2 \cdot L(1+k)}$

$$\begin{cases} (\Delta i_1)_{D_1} = \Delta i_0 \cdot |f_k(\alpha_1, \alpha_2)| \cdot 4(1-\alpha_2) \\ \\ (\Delta i_1)_{D_3} = \Delta i_0 \cdot |1 - f_k(\alpha_1, \alpha_2)| \cdot 4\alpha_1 \end{cases} \quad (13)$$

$$\Delta i_1 = max\left(\Delta i_1|_{D_1}, \Delta i_1|_{D_3}\right) \quad (14)$$

Finally, in the $\alpha_1 < \alpha_2$ case, CRR in first winding is computed as

$$CRR = \frac{\Delta i_1}{\Delta i_0} = max\begin{pmatrix} |f_k(\alpha_1, \alpha_2)| \cdot 4(1-\alpha_2) \\ |1 - f_k(\alpha_1, \alpha_2)| \cdot 4\alpha_1 \end{pmatrix} \quad (15)$$

and when $\alpha_1 > \alpha_2$, with

$$CRR = \frac{\Delta i_1}{\Delta i_0} = max\begin{pmatrix} |f_k(\alpha_1, \alpha_2)| \cdot 4(1-\alpha_1) \\ |1 - f_k(\alpha_1, \alpha_2)| \cdot 4\alpha_2 \end{pmatrix} \quad (16)$$

CRR in the first winding is easily calculated for any combination of duty-cycles and the results are shown as a color map in Figure 7 for $k = 0.9$. Black dotted line shows a current ripple increase of +10%. The closer this line is from diagonal $\alpha_1 = \alpha_2$ the more the structure is constrained in terms of duty-cycle differences. Harsh constrains appear around $\alpha_1 = 0.50$ (Figure 7). Current ripple rise is higher than 10% if duty-cycle difference reaches 0.005. The current controls have to manage to limit the duty-cycles difference between each inverter below this critical value. Ensuring that the duty-cycles difference remains below 0.005 enables to safely exploit the studied architecture degrees of freedom and limit the additional current ripple below the 10% chosen limit. This 0.005 value is consistent with an at least 8-bit duty-cycle quantification. Indeed, duty-cycle differences can be used to equalise both low-frequency currents components without inducing more than a 10% rise of high-frequency current ripple. Purely inductive model gives a pertinent estimation of delay and duty-cycle consequences. It also provides an analytical expression of the current ripple rise induced by winding subdivision concept. Technological delay does not induce more than 10% rise of current ripple. Under the same limit, duty-cycle can be used to balance average currents in subdivided windings as the duty-cycle quantum is under the maximum duty-cycle difference. Nonetheless, this model omits several phenomena, such as parasitic capacitance or losses in conductors and magnetic materials. In order to compute high-frequency losses and improve current estimation, another model has to be considered. As explained in Section 1, finite elements methods require high compute time consumption, therefore, frequency resolution is considered.

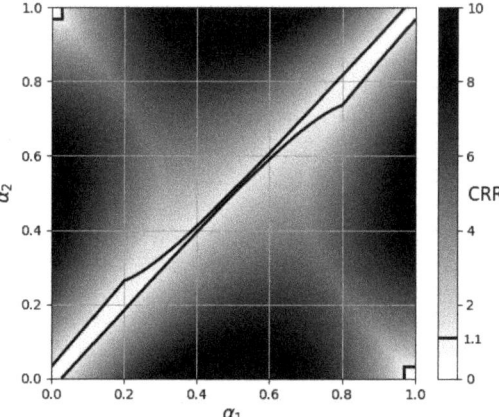

Figure 7. Color map of current ripple ratio CRR for different duty-cycle values α_1 and α_2 under $k = 0.9$. Black line shows 10% current ripple rise limit.

3. Current Ripple Estimation from Admittance Measurements

3.1. Current Harmonics Computation

In previous section, the basic inductive model represented in Figure 4b is employed through (3) in previous section but it can also be represented in a spectral way through its admittance matrix:

$$\underline{Y} = \begin{pmatrix} \dfrac{1}{j(1-k^2)L\omega} & \dfrac{-k}{j(1-k^2)L\omega} \\ \dfrac{-k}{j(1-k^2)L\omega} & \dfrac{1}{j(1-k^2)L\omega} \end{pmatrix} \quad (17)$$

Technically, two autonomous inverters provide a PWM voltage to each subdivided windings. The voltages u_1 and u_2 represented in Figure 5b or Figure 6 are periodic with a switching frequency $F_s = 25$ kHz. Considering each n order harmonic of these voltages $\underline{U}_{1n}$ and $\underline{U}_{2n}$ enables to compute a spectral estimation of each n order current harmonic $\hat{\underline{I}}_{1n}$ and $\hat{\underline{I}}_{2n}$ according to the admittance matrix (17). Indeed, the admittance matrix of the n order pulsation ω_n permits to link directly the n order voltage harmonic to the n order current one:

$$\begin{pmatrix} \hat{\underline{I}}_{1n} \\ \hat{\underline{I}}_{2n} \end{pmatrix} = \begin{pmatrix} \underline{Y}_{11}(\omega_n) & \underline{Y}_{12}(\omega_n) \\ \underline{Y}_{21}(\omega_n) & \underline{Y}_{22}(\omega_n) \end{pmatrix} \cdot \begin{pmatrix} \underline{U}_{1n} \\ \underline{U}_{2n} \end{pmatrix} \quad (18)$$

Finally, the subdivided windings currents can be estimated by adding each harmonic contribution. The admittance matrix in (17) corresponds to purely inductive model. This model neglects the resistive behaviour of conductors at switching frequency range, among other things. Therefore, it has to be supplemented by an electrical characterisation. The aim of the following subsection is to establish this matrix on admittance measurements conducted on a practical device in order to refine knowledge of actual windings electrical behaviour. The inductive model and the one based on the admittance measurements are compared through the same method detailed in (18).

3.2. Admittance Matrix Measurements

In (18), two types of terms have to be detailed. The diagonal terms can be directly measured with a 1-port Impedance Analyser as, for example $\underline{Y}_{11} = (\underline{I}_1/\underline{U}_1)_{\underline{U}_2=0}$ requires current and voltage on the same winding. Nevertheless, trans-admittance terms like $\underline{Y}_{12} = (\underline{I}_1/\underline{U}_2)_{\underline{U}_1=0}$ cannot be provided by this measuring instrument as current and voltage are not measured in the same winding. In this part, possible measures are detailed and then, the derived computation of trans-admittance terms is also explained.

A practical device is built in order to test a proof of concept and validate subdivided windings modelling detailed in Section 2. This device is composed of two 20-turn windings wound around a laminated steel magnetic circuit without airgap as shown in Figure 8a. Lack of airgap moves away from the electric machine context but it provides two windings with similar electrical properties at switching frequency range enabling to exclusively focus on how voltages time delays and duty-cycles differences may impact sub-coils currents ripples. Once the practical device had been created, small signals impedance measurements are carried out using a Keysight E4990A Impedance analyser. These are convenient, harmless and provide information on the device electrical properties over a wide frequency range. For each winding p, while other winding q is opened $\left(\underline{Y}_p\right)_{i_q=0} = \left(\underline{I}_p/\underline{U}_p\right)_{I_q=0}$ (dotted lines in Figure 9a) or short-circuited $\underline{Y}_{pp} = \left(\underline{I}_p/\underline{U}_p\right)_{\underline{U}_q=0}$, 1600 admittance measures are performed from the switching frequency $F_s = 25$ kHz up to 40 MHz corresponding to the one thousand six hundredth harmonic.

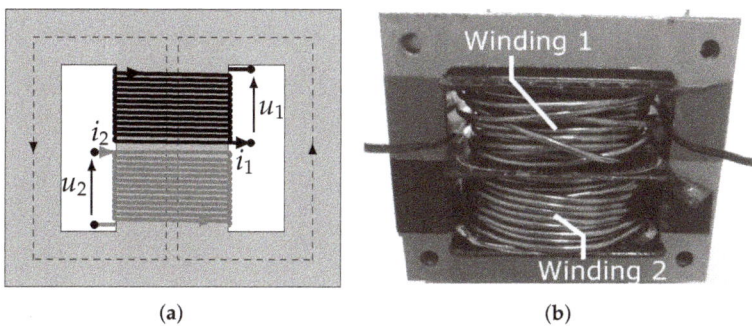

Figure 8. Symmetrical model with 20-turns windings around magnetic circuit without airgap. (**a**) Theoretical model, (**b**) Practical device.

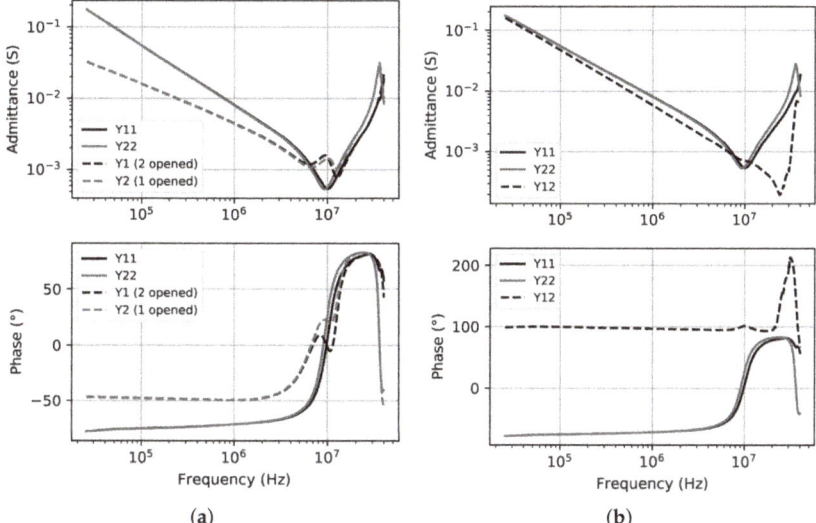

Figure 9. (**a**) Admittance measurements leading to (**b**) the terms of admittance matrix for 20-turns windings around magnetic circuit without airgap.

Figure 9a shows that both windings are nearly identical for frequency under 5 MHz. Thus, it appears that the symmetry hypothesis is verified for this practical device. Nonetheless, cross-coupling terms cannot be directly measured. Based on the real measurements (Figure 9a), only the product of $\underline{Y}_{12}$ and $\underline{Y}_{21}$ can be estimated with any of the following equivalent relations:

$$(\underline{Y}_{12} \cdot \underline{Y}_{21})_1 = \left[\underline{Y}_{11} - (\underline{Y}_1)_{i_2=0}\right] \cdot \underline{Y}_{22} \quad \text{or} \quad (\underline{Y}_{12} \cdot \underline{Y}_{21})_2 = \left[\underline{Y}_{22} - (\underline{Y}_2)_{i_1=0}\right] \cdot \underline{Y}_{11} \quad (19)$$

As admittance matrix is symmetrical, it derives

$$\underline{Y}_{12} = \underline{Y}_{21} = \pm\sqrt{\langle \underline{Y}_{12} \cdot \underline{Y}_{21}\rangle_{1,2}} \quad (20)$$

Sign of $\underline{Y}_{12} = \underline{Y}_{21}$ is chosen with respect to coupling sign convention represented in Figure 4b. Note that direct trans-admittance measurements using Network Analyser and current probe can also be considered as in [23] for non-symmetrical matrices.

Figure 9b shows the admittance matrix terms computated for coupled inductors sharing the same laminated circuit without any airgap as described in Figure 8a. This term contains information about inductance value dispersion on a wide frequency range. A resonance appears on both windings around 10 MHz indicating a parasitic capacitance of sub-coils. As this measurements are used for current ripple estimation and power loss computing, only values corresponding to harmonic of a lower order than 200 (i.e., 5 MHz) are selected in order to guarantee symmetry hypothesis and to simplify the computation. Indeed, main part of losses is due to current harmonic under 1 MHz with an uncertainty of 0.1%.

3.3. Comparison of Estimated Current Shape

The measurements presented above are now used to estimate the current ripple during a switching period. The related results are compared with the compared ones provided by the alternative basic inductive model. Figure 10 shows in dashed lines the current shapes obtained with the inductive model described in Figure 4b. The inductive model is based on inductance and coupling factor values measured at switching frequency $F_s = 25$ kHz : $L = 190$ µH and $k = 0.91$. These current shapes are compared to the estimated ones using all harmonics contribution between 50 kHz to 1 MHz according to (18).

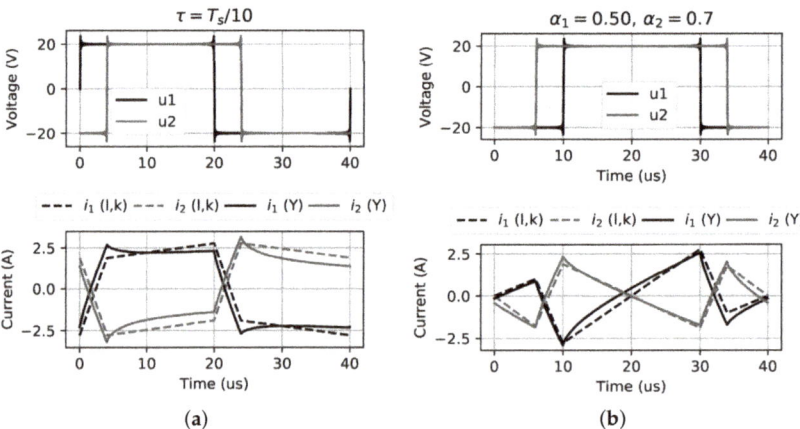

Figure 10. Comparison between purely inductive and admittance (Y) based models, $F_s = 25$ kHz. (**a**) Delay case, (**b**) Duty-cycle difference case.

In both current shapes presented in Figure 10a, the chosen delay is not realistic compared to the practical desynchronisation that may occur using modern technologies (i.e., $\tau = 50$ ns). Nevertheless, this delay shows significant impacts on the current ripple and permits to easily compare both models. Specifically, a power flow from the first to the second sub-windings illustrates useless high-frequency additional losses that the proposed architecture must face. It appears through the fact that i_2 is opposed to u_2 whereas u_1 and i_1 are in phase. This experimental result confirms the expected phenomenon presented in Figure 5b using the theoretical inductive model (Section 2.2). Figure 10b validates also currents waveforms predicted by inductive model in case of voltages duty-cycles differences among applied voltages. In both configurations, namely time delay or duty-cycles differences, the estimated current waveform based on admittance model is linear during differential mode, corresponding to $u_1 = -u_2$, and follows exponential branches during common mode, corresponding to $u_1 = u_2$.

Because of positive coupling between both sub-windings, the equivalent inductance is low during the differential mode phase leading to fast linear current evolution. On the other hand, during common mode phases, the current evolution is slower and follows exponential branches. In this common mode phases, sub-windings present a higher inductance than differential mode whereas conductor resistance is unchanged.

These comments support the findings described using the basic inductive model while offering more precise information on the actual current waveform. This current ripple estimation based on the admittance measurements also permits to compute extra-losses due to the voltages differences. The estimated losses are then interpreted in the next section through a comparison with measured losses.

4. PWM-Induced Current Ripple: Experimental Validation and Losses Estimation

4.1. Experimental Setup

In order to validate the proposed current estimation method, an experimental system is set up. This experimental device (Figure 8b) consists in two four-quadrant fast switching IGBT inverters which independently supply two-coils wound around the same magnetic core. Both inverters are connected to the same DC low voltage bus (Figure 11). An Arduino card provides control signals for both inverters in order to supply the two sub-windings with voltages similar to the theoretical waveforms presented in Figure 5b or Figure 6. The switching frequency is set to 25 kHz and the time resolution of the delay between both control signals is 0.1 μs. As far as voltages u_1 and u_2 are concerned, the microcontroller enables to control their time delay and their duty-cycles difference. The low DC-link voltage is provided by a stabilised power supply. Its value is chosen considering that, in electric machine context, a 1T induction varying at 500 Hz in the chosen magnetic core would induce a 20 V EMF in the 20-turn sub-windings. Therefore, the power supply regulates DC-link voltage to a 20 V value. Two current sensors measure each sub-winding current. The voltage is also measured at each winding terminal. The four resulting signals are sampled at 20 MHz by an oscilloscope and processed using a Python routine.

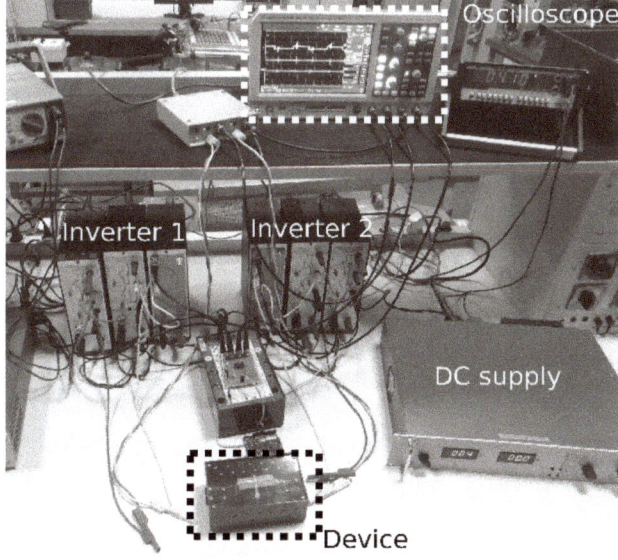

Figure 11. Experimental Setup.

First, the inverters feeding both windings are synchronised. Voltages and currents measured on the device under test are shown in Figure 12. The harmonics of the measured voltages are used in (18) to estimate the sub-windings currents which requires a precise knowledge of the admittance matrix. This measured current waveforms have similar shapes than the ones predicted using the frequency model (Section 2) but with lower magnitude. This can be explained by the fact that the admittance measurements are carried out with a impedance meter and hence using low voltage values, namely a voltage magnitude forty times lower than the one generated by the PWM inverters. Non-linearity causes the admittance matrix to depend on the voltage magnitude and must explain observed deviations.

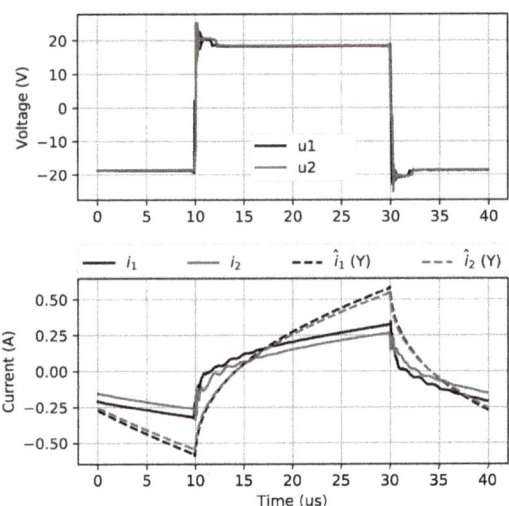

Figure 12. Current and voltage measurements with synchronous control signals. The dotted line shows the estimate current waveforms.

4.2. Delay Study

After having validated the contrast between estimated and measured currents in a synchronised case, time delay effect is investigated. Section 2.2 has shown a transfer from the winding whose voltage is in phase advance with respect to the other. This phenomenon is particularly noticeable in practical experiment since time delays are deliberately set to a much higher than the expected ones (Section 2). A symmetrical configuration represented by the practical device without airgap is tested. In case where u_2 presents a 2 µs delay with u_1, has shown in Figure 13a, power flowing from winding 1 to winding 2 is also visible on the magnitude of each voltage. Although, the DC-link voltage is regulated to 20 V, u_2 presents a higher continuous value (21–22 V) whereas u_1 maximum continuous value is slightly lower than 20 V. This imbalance in DC voltage inverters inputs values is a consequence of the described adverse power flow; it is a consequence of the actual unavoidable resistive connections of the DC-bus. The use of a symmetrical model proves that this imbalance is only due to the delay between voltages.

Figure 13a,b show that during the phase where $u_1 = -u_2$, voltages measured at the winding terminals drops by 5 V compared to the DC-bus voltage because connections impedances are no longer negligible compared to the device impedance. This is so because the coupled inductors have low inductance in differential mode as shown in Table 1. At 25 kHz, the device under test presents a 20 µH inductance which is almost in same order of magnitude than connections parasitic inductance.

This effect tends to minimise the impact of the delay on the current ripple rise compared to the theoretical predictions (1) but it is taken into account in the above estimation.

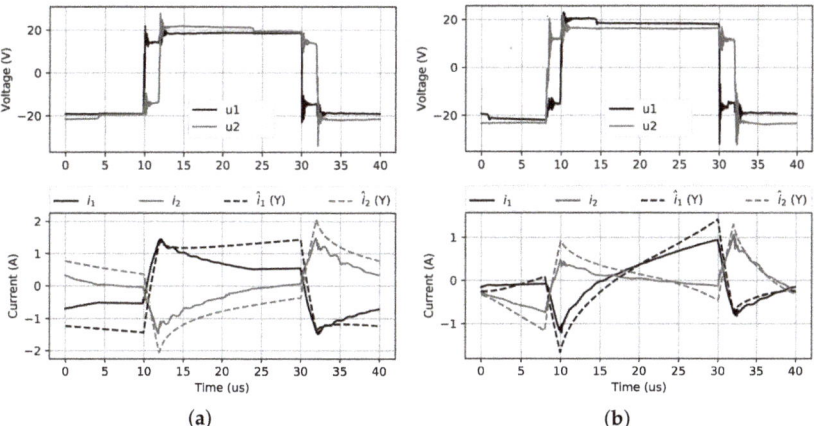

Figure 13. Current and voltage measurements with (a) 2 μs delayed control signals, (b) $\alpha_1 = 0.50$ and $\alpha_2 = 0.60$.

4.3. Duty-Cycle Discrepancies

Admitting that the control signals are now synchronised, the duty-cycles difference is henceforth investigated. The configuration for which α_2 is higher than α_1 is presented in Figure 13b. The average value of u_2 is positive, so i_2 evolves around a continuous average value of 3 A whereas i_1 is centred around 0 A. The average value of i_2 is suppressed in order to only visualise its current ripple. The estimated and measured currents are depicted in Figure 13b; they show some disparities but current shapes are similar in both differential and common modes. The windings currents average values induce a magnetic circuit polarisation that must change admittance matrix terms. A thermal drift also occurs when continuous average currents are maintained which subsequently also alter these terms. These phenomena are not taken into account in the present study. These aspects introduce some inconsistencies between estimations and measurements.

4.4. Losses Estimation

Based on the current and voltage measurements, losses $P_{measure}$ occurring within experimental device are calculated as showed in (21).

$$P_{measure} = \frac{1}{T_s} \int_0^{T_s} (u_1(t) \cdot i_1(t) + u_2(t) \cdot i_2(t)) \, dt \qquad (21)$$

It can be compared to $P_{estimation}$ (22) resulting by adding up all the individual harmonic losses estimated from the admittance measurements and the spectral decomposition of the measured voltages.

$$P_{estimation} = \Re \left(\sum_{n=1}^{\infty} \left[\underline{U}_{1n} \cdot \underline{\hat{I}}_{1n}^* + \underline{U}_{2n} \cdot \underline{\hat{I}}_{2n}^* \right] \right) \qquad (22)$$

where $\underline{U}_{pn}$ is the n-order complex voltage harmonic from u_p spectral decomposition measured at winding p terminals and $\underline{\hat{I}}_{pn}$ is the n-order complex current harmonic estimation in the sub-coil p.

The losses based on the measured electrical variables are compared with the estimated losses computed by injecting the PWM voltage waveform in the frequency model. The comparative results are shown in Figure 14. In the same way as in Section 3.3, the chosen delay is not realistic compared to the practical desynchronisation that may occur using modern technologies. The expected technological delay is shown as a light grey area in Figure 14a, by using the value calculated in accordance with the methodology lai down in Section 2.2. In the case of duty-cycle differences, the comparison is made for duty-cycle differences values that are higher than the limit fixed in Section 2.3. This limit is represented by a light grey area in Figure 14b and corresponds to the degrees of freedom range that can be used to balance the low-frequency currents in each sub-windings. In any case, the estimated losses are roughly 25% higher than the measured ones. As previously explained, these inconsistencies must be related to non-inclusion of thermal aspect and magnetic circuit polarisation phenomenon in the proposed estimation method. System non-linearity causes low voltage admittance measurements to differ from its actual value. Finally, losses comparison shows that the estimation based on the admittance measurements can be corrected by a proper normalisation. By multiplying admittance magnitude by 0.7, the corrected estimation losses are close to measured losses. With this correction factor, the model based on the admittance measures provides a reliable and effective tool for current ripple estimation and high-frequency power losses assessment in the context of subdivided windings.

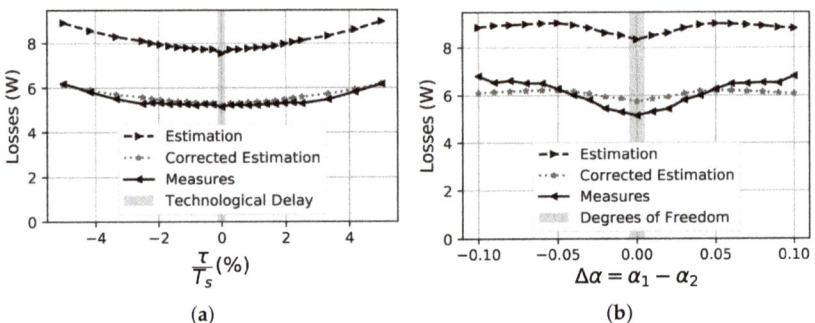

Figure 14. Estimated, Estimated after correction and Measured Losses under (**a**) delay and (**b**) duty-cycle difference.

5. Discussion

The concept of subdividing stator winding [14] is presented in Section 1. On the basis of this innovative principle, the DC-link voltage can be significantly lowered and the combination of several inverters with their relative sub-windings provides a more even PWM voltage distribution than in the classic single-winding–single-inverter combination. Hence, the new configuration should largely reduce the ageing of machine dielectric insulating materials which are extremely impacted at the coil ends by even distribution of dv/dt in standard architectures. Moreover lower voltages throughout the electric drive enables safer maintenance operations and reduces integration constraints.

In addition to these positive effects, the novel studied architecture offers new degrees of freedom since each sub-winding can now be independently controlled. Considering two sub-coils located in the same stator slot, this degree of freedom has to be carefully managed since, by design, the two adjacent coils are highly magnetically coupled. The present study examines the new requirements for the inverter and its associated control system to ensure that they respect these physical constraints. A high degree of control precision is required, which is particularly difficult to achieve considering a single switching period. All other things being equal, the current study demonstrates that actual technological devices permit to follow the essential requirements and allow to operate the new architecture safely.

Obviously, the studied architecture distributes the global power to several small inverter-sub-coil combinations and consequently leads to consider wide-bandgap techology switches such as GaN transistors. They allow to use higher switching frequencies and generate higher dv/dt. Consequently, regarding these new parameters, the problem should be reconsidered on the basis of the present methodology. Addressing this more general issue will enable to tackle the global optimisation of the studied architecture, that is to determine the three optimal parameters, namely the number of fractionation n, the PWM frequency Fs and the power switch technology.

This perspective clearly shows that the reported work is a necessary step to ensure the technological feasibility of this subdivided structure. It provides a good insight into the key parameters driving the magnetic interactions in the subdivided windings at the critical switching frequency range. To support the findings, a proof of concept is designed, implemented and extensively detailed. Some of the system parameters are deliberately fixed. To scale up to a prototype level and address the entire machine drive context, it is necessary to consider all possible parameters, which is the next step.

6. Conclusions

In the case of two subdivided windings, the sub-coil voltage constraints are studied using an inductive model. With respect to this specific sub-coil and sub-inverter combination, the main issue is related to any voltage disparities over a PWM switching period. In this case, the two key parameters are the time offset and the duty-cycle difference between the two sub-coil voltages. Both parameters greatly impact the currents in the sub-windings. Subsequently, the model is used to assess the currents waveforms and to evaluate the related current ripple which is compared to that of a standard winding-inverter topology. A theoretical approach enables to compute the maximum operating range of both parameters in order to analyse the technical viability of the studied structure compared to the classical one. The admissible range of the time delay between both sub-coils voltages is consistent with current technologies used to drive power switches. The permissible duty-cycles difference range is also compatible with the range of variation for balancing the low-frequency currents in the subdivided windings. To further reinforce these theoretical results, the understanding of electrical phenomena in sub-windings is improved through admittance measures providing information on its actual electrical behaviour. The resulting method provides an accurate estimate of the current ripple and also the related power losses. This estimation tool is validated by testing a proof of concept system combining two subdivided coils wound around the same magnetic core and supplied by two independent inverters with a common DC bus. The series of various voltage tests confirms the theoretical findings. In most cases, the proposed method of the current waveform estimation provides a reliable model to represent interaction between sub-windings and will help to give a good insight of the winding subdivision concept. However the symmetrical modelling of subdivided windings does not represent properly the effect of the air-gap in the context of an electrical machine. To take the analysis one step further, the ongoing work is to extend the present study to the case of asymmetrical windings.

7. Patents

Patent WO2018149996 (https://patentscope.wipo.int/search/fr/detail.jsf?docId=WO2018149996).

Author Contributions: Conceptualization, E.L., J.O. and O.B.; methodology, all authors; software, A.C.; validation, all authors; formal analysis, A.C.; resources, J.O.; investigation, A.C.; visualisation, A.C.; writing–original draft preparation, A.C.; writing–review and editing, all authors; supervision, O.B., E.L. and J.O.; project administration, J.O. and O.B.; funding acquisition, O.B. and J.O.

Funding: This research was funded by Ministère de l'Enseignement Supérieur, de la Recherche et de l'Innovation of France. The patent is supported by Centre National de la Recherche Scientifique of France. A research project related to this patent is scientifically and financially supported by both laboratories GeePs and SATIE.

Acknowledgments: The authors would like to thank the members of the technical support team for their help in the realisation of practical models.

Conflicts of Interest: The authors declare no conflict of interest.

References

1. Cissé, K.M.; Hlioui, S.; Cheng, Y.; Belhadi, M.H. Etat de l'art des topologies de machines électriques utilisées dans les véhicules électriques et hybrides. In Proceedings of the Symposium de Genie Electrique (Sge 2018), Nancy, France, 3–5 July 2018; pp. 3–5.
2. Oswald, N.; Anthony, P.; McNeill, N.; Stark, B.H. An experimental investigation of the tradeoff between switching losses and EMI generation with hard-switched All-Si, Si-SiC, and All-SiC device combinations. *IEEE Trans. Power Electron.* **2014**, *29*, 2393–2407. [CrossRef]
3. Han, D.; Li, S.; Lee, W.; Choi, W.; Sarlioglu, B. Trade-off between switching loss and common mode EMI generation of GaN devices-analysis and solution. In Proceedings of the Conference Proceedings—IEEE Applied Power Electronics Conference and Exposition—APEC, Tampa, FL, USA, 26–30 March 2017; pp. 843–847,[CrossRef]
4. Concari, L.; Barater, D.; Toscani, A.; Concari, C.; Franceschini, G.; Buticchi, G.; Liserre, M.; Zhang, H. Assessment of efficiency and reliability of wide band-gap based H8 inverter in electric vehicle applications. *Energies* **2019**, *12*, 1922. [CrossRef]
5. Zoeller, C.; Wolbank, T.M.; Vogelsberger, M.A. Inverter-fed drive stator insulation monitoring based on reflection phenomena stimulated by voltage step excitation. In Proceedings of the ECCE 2016 IEEE Energy Conversion Congress and Exposition, Milwaukee, WI, USA, 18–22 September 2016; pp. 1–8. [CrossRef]
6. Florkowski, M.; Florkowska, B.; Zydron, P. Partial discharges in insulating systems of low voltage electric motors fed by power electronics—Twisted-pair samples evaluation. *Energies* **2019**, *12*, 768. [CrossRef]
7. Erdman, J.M.; Kerkman, R.J.; Schlegel, D.W.; Skibinski, G.L. Effect of PWM Inverters on AC Motor Bearing Currents and. *IEEE Trans. Ind. Appl.* **1996**, *32*, 250–259. [CrossRef]
8. Wang, J.; Li, Y.; Han, Y. Integrated Modular Motor Drive Design WithGaNPowerFETs. *IEEE Trans. Ind. Appl.* **2015**, *51*, 3198–3207. [CrossRef]
9. Gerrits, T.; Wijnands, C.G.E.; Paulides, J.J.H.; Duarte, J.L. Electrical gearbox equivalent by means of dynamic machine operation. In Proceedings of the 2011 14th European Conference on Power Electronics and Applications, Birmingham, UK, 30 August–1 September 2011; pp. 1–10.
10. Hoang, E.; Gaussens, B.; Lecrivain, M.; Gabsi, M. Proposition pour accroître la puissance convertible par un ensemble onduleur de tension—Machine synchrone à commutation de flux à double excitation dans une application motorisation de véhicule hybride ou électrique. In Proceedings of the Symposium de Genie Electrique (Sge"14), Cachan, France, 8–10 July 2014; pp. 8–10.
11. Li, G.; Ojeda, J.; Hoang, E.; Gabsi, M. Double and single layers flux-switching permanent magnet motors: Fault tolerant model for critical applications. In Proceedings of the 2011 International Conference on Electrical Machines and Systems, ICEMS 2011, Beijing, China, 20–23 August 2011; [CrossRef]
12. Li, W.; Cheng, M. Reliability analysis and evaluation for flux-switching permanent magnet machine. *IEEE Trans. Ind. Electron.* **2019**, *66*, 1760–1769. [CrossRef]
13. Baumgardt, A.; Bachheibl, F.; Patzak, A.; Gerling, D.; Machine, A.E. 48V Traction: Innovative Drive Topology and Battery. In Proceedings of the 2016 IEEE International Conference on Power Electronics, Drives and Energy Systems (PEDES), Trivandrum, India, 14–17 December 2016. [CrossRef]
14. Hoang, E.; Labouré, E. Electric Machine Powered at Low Voltage and Associated Multicellular Traction Chain. WO2018149996, 23 August 2018.
15. Boxriker, M.; Kolb, J.; Doppelbauer, M. Expanding the operating range of permanent magnet synchronous motors by using the optimum number of phases. In Proceedings of the 2016 18th European Conference on Power Electronics and Applications, EPE 2016 ECCE Europe, Karlsruhe, Germany, 5–9 September 2016; pp. 1–8. [CrossRef]
16. Ojeda, J.; Bouker, H.; Vido, L.; Ahmed, H.B.; Cachan, E.N.S. Comparison of 3-phase and 5-phase high speed synchronous motor for EV/HEV applications. In Proceedings of the 8th IET International Conference on Power Electronics, Machines and Drives (PEMD 2016), Glasgow, UK, 19–21 April 2016; [CrossRef]
17. Wu, B.; Xu, D.; Ji, J.; Zhao, W.; Jiang, Q. Field-Oriented Control and Direct Torque Control for a Five-Phase Fault-Tolerant Flux-Switching Permanent-Magnet Motor. *Chin. J. Electr. Eng.* **2018**, *4*, 48–56. [CrossRef]

18. Bojoi, R.; Rubino, S.; Tenconi, A.; Vaschetto, S. Multiphase electrical machines and drives: A viable solution for energy generation and transportation electrification. In Proceedings of the 2016 International Conference and Exposition on Electrical and Power Engineering, EPE 2016, Iasi, Romania, 20–22 October 2016; pp. 632–639. [CrossRef]
19. Gubbala, L.; Von Jouanne, A.; Enjeti, P.; Singh, C.; Toliyat, H. Voltage distribution in the windings of an AC motor subjected to high dV/dt PWM voltages. In Proceedings of the PESC Record—IEEE Annual Power Electronics Specialists Conference, Atlanta, GA, USA, 18–22 June 1995; Volume 1, pp. 579–585. [CrossRef]
20. Ghassemi, M. Accelerated insulation aging due to fast, repetitive voltages: A review identifying challenges and future research needs. *IEEE Trans. Dielectr. Electr. Insul.* **2019**, *26*, 1558–1568. [CrossRef]
21. Chang, L.; Jahns, T.M. Prediction and Evaluation of PWM-Induced Current Ripple in IPM Machines Incorporating Slotting, Saturation, and Cross-Coupling Effects. *IEEE Trans. Ind. Appl.* **2018**, *54*, 6015–6026. [CrossRef]
22. Tamizi, K.; Béthoux, O.; Labouré, E. An easy to implement and robust design control method dedicated to multi-cell converters using inter cell transformers. *Math. Comput. Simul.* **2019**, [CrossRef]
23. Gustavsen, B. Wide band modeling of power transformers. In Proceedings of the 2004 IEEE Power Engineering Society General Meeting, Denver, CO, USA, 6–10 June 2004; Volume 2, p. 1791.

© 2019 by the authors. Licensee MDPI, Basel, Switzerland. This article is an open access article distributed under the terms and conditions of the Creative Commons Attribution (CC BY) license (http://creativecommons.org/licenses/by/4.0/).

Article

A Generalised Multifrequency PWM Strategy for Dual Three-Phase Voltage Source Converters

Jose A. Riveros [1,2,*], Joel Prieto [3], Marco Rivera [1], Sergio Toledo [1,4] and Raúl Gregor [4]

1. Facultad de Ingeniería, Universidad de Talca, Curicó 3341717, Chile; marcoriv@utalca.cl (M.R.); stoledo@utalca.cl (S.T.)
2. Facultad Politécnica, Universidad Nacional de Asunción, San Lorenzo 2111, Paraguay
3. Facultad de Ciencias de la Ingeniería, Universidad Paraguayo Alemana, San Lorenzo 2160, Paraguay; joel.prieto@upa.edu.py
4. Facultad de Ingeniería, Universidad Nacional de Asunción, Luque 2060, Paraguay; rgregor@ing.una.py
* Correspondence: jriveros@utalca.cl; Tel.: +56-752-315-470

Received: 23 March 2019; Accepted: 9 April 2019; Published: 11 April 2019

Abstract: Pulse width modulation (PWM) strategies for the control of asymmetrical six-phase drives have been widely studied since the beginning of this century. Nevertheless, space vector modulation (SVM) techniques with multifrequency voltage injection for the control of all the degrees of freedom of the multiphase model is still a subject under research. This paper deals with this topic and introduces a generalised PWM method for a two-level voltage source converters. The architecture was derived by extending a three-phase modulator proposed as an alternative to the widely studied SVM. The proposal computes the duty times straightforwardly with a fast algorithm based on an analytical solution of the voltage-time modulation law. Theoretical derivations supported by experimental results demonstrate the proper synthesis of the multifrequency target voltage in the linear modulation region as well as good frequency behaviour of the presented modulation strategy.

Keywords: multiphase drives; pulse width modulation; dc-ac power converters

1. Introduction

Multiphase technology has become one of the most attractive subjects within the electric drives research area [1]. Since the beginning of this century, numerous publications in journals and dedicated sessions of conferences have reported innovative exploiting of their additional number of degrees of freedom respect to the conventional topology [2]. Thus, high-performance control strategies have been developed to take advantage of the most promoted features such as fault-tolerant, efficient electromechanical energy conversion and distribution of the current stress into more than three phases [3–5]. These are commonly designed to be used in applications such as electric vehicles, ship propulsion, renewable energy generation and high-power industry [6]. The development of modulation and control schemes capable of regulating the entire multiphase modelling, compounded by multiple two-dimensional subspaces [7], is still one of the main topics [3,4]. Implementations with low computational cost are the desired and most impacting result in this task, considering that the hardware barriers (digital controllers, topologies and power switches) have been overcome with the last advancements and maturity achieved in other involved fields [2–4].

The asymmetrical dual three-phase machines are one of the most considered designs within multiphase systems [1]. This proposal is compounded by two sets of three-phase windings electrically shifted by 30 degrees in order to attain the best torque ripple behaviour. This topology presents compatibility with the three-phase power converters available in the market, which can be associated to become a six-phase voltage source converter (VSC) and supply the described multiphase machine. Moreover, the sets of windings can be connected with single or double neutral-point formats. The first configuration provides enhancements in the electromagnetic torque generation [8] and better post-fault characteristics [9]. On the other hand, the isolated neutrals arrangement enables the best dc-bus utilisation [10] and less susceptibility to the low-order stator current harmonics (the triplen components are annulled) [11]. This proposal currently meets the requirements to replace the three-phase drives in high-power applications. Nevertheless, new advancements in the control and modulation strategies will contribute to the gradual adoption of the multiphase technology in non-conventional uses.

The pulse width modulation (PWM) methods for the asymmetrical dual three-phase drives have been widely covered in different researches [12–16]. These techniques are developed by using the vector space decomposition (VSD) approach [12]. Thus, the multiphase inverter modelling is represented in multiple complex planes. The main subspace held the fundamental frequency component, whereas the harmonics and zero-sequence variables are located in the complementary planes. The resulting space can be organised in twelve [13] or twenty-four [14] sectors to develop the voltage space vector modulation (SVM) strategy. A higher number of sectors enhances the performance in terms of flux harmonic distortion factor at the expense of a higher computational burden [17]. For the overmodulation region, the implementation of [15] employs two three-phase SVM with a competitive classification algorithm, and a minimum harmonic distortion modulator was presented in [16]. All the reviewed proposals had been developed considering a single frequency reference voltage (zero command voltage in the harmonics' subspace).

The surveyed schemes do not provide the highest performance in real multiphase drives, because they lead unwanted low-order stator current harmonics [18]. These are caused by the small constructive asymmetries of the electric machine and, in higher proportion, by the non-linear effects of the switching dead-time. Injection of current in the complementary subspaces (possible with references harmonics voltage different than zero) of the multiphase modelling is required to overcome this drawback. For this reason, multifrequency modulation schemes are an interesting research topic in this field. However, this subject has been barely covered in the literature considering the six-phase VSCs and implementations report the use of double zero-sequence injection with the carrier-based approach to accomplish with this goal [2]. Recently, two proposals were assessed in [19,20] as an extension of the five-phase modulator presented in [21]. Two voltage-time equation systems (one for the fundamental frequency and another for the harmonics' plane) are solved, and a time-multiplexing has been used for the application of these solutions within a sampling period. The technique could employ up to eight active voltage space vectors (twice the necessary number to control both planes of the model) to synthesise independent voltage outputs at two different frequencies. Additionally, an asymmetric switching pattern (characterised by higher current ripples and more complex implementation) was reported in the validations of this modulation architectures with simulation results.

A new PWM method is introduced in this work. The strategy controls all the degrees of freedom of the six-phase VSC by employing a generalised analytical solution of the modulation law. The technique is an extension of the three-phase modulator recently developed in [22]. This approach has been extrapolated by adding the harmonics' subspace with the VSD theory and arranging the model to attain two decoupled three-phase modulators. These last are commanded by two auxiliary voltages defined straightforwardly from the original reference voltage vectors. The configuration allows the operation in the multifrequency mode with a low computational cost algorithm. The inputs of the architecture are the reference components in the stationary reference frame and the zero-sequence control signal to compute the duty cycles of the inverter's legs. Consequently, the magnitude and position (sector) are not necessary, avoiding the high computational effort required for operators such as squared root or trigonometric functions. Continuous switching and healthy operation of the inverter within the linear modulation interval is the scope for the introduction of the proposal.

This paper is organised as follows. The next section reviews the modelling of the system with emphasis in the output voltages. Then, the developed modulation strategy is detailed in Section 3. Next, Section 4 discusses the experimental validation for the time- and frequency-domain. The last part presents the conclusions obtained after the evaluation process.

2. Modelling of the Dual Three-Phase Voltage Source Converter

The two-level six-phase VSC is a popular topology within the research area because of their promising features [4]. An electrical diagram of this design is shown in Figure 1. The inverter is power-supplied through the dc-bus with a V_{dc} input voltage. This is processed by means of an arrangement of six legs, which in turns are composed by two power semiconductors in series. These last must operate in the complementary conduction mode to avoid damaging currents. Thus, the switching state S_j ($j = \{a, b, \ldots, f\}$) can be modelled with a bit signal, where $S_j = 1(0)$ indicates that the top(bottom) power switch of the leg j is activated. The variables of the three-phase sets are designated with the subscripts a-b-c and d-e-f, respectively. The switching functions provide different phase voltages V_j, and a total of 64 combinations can be generated and calculated as follows [11]:

$$\begin{bmatrix} V_a \\ V_d \\ V_b \\ V_e \\ V_c \\ V_f \end{bmatrix} = \frac{V_{dc}}{3} \begin{bmatrix} 2 & 0 & -1 & 0 & -1 & 0 \\ 0 & 2 & 0 & -1 & 0 & -1 \\ -1 & 0 & 2 & 0 & -1 & 0 \\ 0 & -1 & 0 & 2 & 0 & -1 \\ -1 & 0 & -1 & 0 & 2 & 0 \\ 0 & -1 & 0 & -1 & 0 & 2 \end{bmatrix} \begin{bmatrix} S_a \\ S_d \\ S_b \\ S_e \\ S_c \\ S_f \end{bmatrix} \quad (1)$$

The scalar model of the VSC derived in (1) can be simplified to facilitate the development of the modulation strategy by applying the VSD approach [12]. The linear transformation with invariant magnitude format is detailed below:

$$\begin{bmatrix} V_\alpha \\ V_\beta \\ V_x \\ V_y \\ V_{0+} \\ V_{0-} \end{bmatrix} = \frac{1}{3} \begin{bmatrix} 1 & C_\delta & -S_\delta & -C_\delta & -S_\delta & 0 \\ 0 & S_\delta & C_\delta & S_\delta & -C_\delta & -1 \\ 1 & -C_\delta & -S_\delta & C_\delta & -S_\delta & 0 \\ 0 & S_\delta & -C_\delta & S_\delta & C_\delta & -1 \\ 1 & 0 & 1 & 0 & 1 & 0 \\ 0 & 1 & 0 & 1 & 0 & 1 \end{bmatrix} \begin{bmatrix} V_a \\ V_d \\ V_b \\ V_e \\ V_c \\ V_f \end{bmatrix} \quad (2)$$

with C_δ and S_δ being the cosine and sine operators of $\delta = \pi/6$, respectively. These are the components in an orthogonal and stationary reference frame. Furthermore, the fundamental frequency along with the $12k \pm 1$ ($k = 1, 2, \ldots$) order harmonics are held in the α-β plane, the $6k \pm 1$ order harmonics are mapped into the x-y subspace and the zero-sequence components are projected in the remaining space. The decoupled components can be also calculated with the switching functions by combining (1)–(2). The result with the normalised voltages respect to $V_b = V_{dc}/2$ is the following:

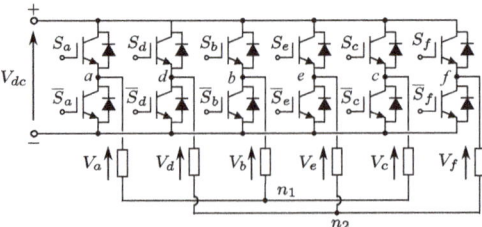

Figure 1. Six-phase voltage source converter (VSC) power-supplying a dual three-phase load with isolated neutrals.

$$\begin{bmatrix} v_\alpha \\ v_\beta \\ v_x \\ v_y \\ v_{0+} \\ v_{0-} \end{bmatrix} = \frac{2}{3} \begin{bmatrix} 1 & C_\delta & -S_\delta & -C_\delta & -S_\delta & 0 \\ 0 & S_\delta & C_\delta & S_\delta & -C_\delta & -1 \\ 1 & -C_\delta & -S_\delta & C_\delta & -S_\delta & 0 \\ 0 & S_\delta & -C_\delta & S_\delta & C_\delta & -1 \\ 0 & 0 & 0 & 0 & 0 & 0 \\ 0 & 0 & 0 & 0 & 0 & 0 \end{bmatrix} \begin{bmatrix} S_a \\ S_d \\ S_b \\ S_e \\ S_c \\ S_f \end{bmatrix} \quad (3)$$

where $v_m = V_m/V_b$ are the normalised voltage components of the m-axis ($m = \{\alpha, \beta, x, y, 0^+, 0^-\}$). This approach cannot be used to study the zero-sequence components, but these are a degree of freedom of the system that could be added later to improve some performance characteristic [22]. For this reason, the 0^+ and 0^- component are omitted from the derivation of the modulation scheme henceforth. The modelling of the VSC is also described by current equations and the common-mode voltage. These are not included for the sake of simplicity in the introduction of the proposed technique.

3. Multifrequency Pulse Width Modulation Algorithm

A modulation space for the dual three-phase VSC described by the switching signals can be derived as an extension of the approach introduced in [22]. Hence, expanding the matrix operators of the modelling (3) derived with the VSD theory, the following result is achieved:

$$\begin{aligned} v_\alpha &= (2/3)(S_a + C_\delta S_d - S_\delta S_b - C_\delta S_e - S_\delta S_c) \\ v_\beta &= (2/3)(S_\delta S_d + C_\delta S_b + S_\delta S_e - C_\delta S_c - S_f) \\ v_x &= (2/3)(S_a - C_\delta S_d - S_\delta S_b + C_\delta S_e - S_\delta S_c) \\ v_y &= (2/3)(S_\delta S_d - C_\delta S_b + S_\delta S_e + C_\delta S_c - S_f) \end{aligned} \quad (4)$$

Then, by following the procedure of [22], the modulation law is achieved by integrating (4) over time within a sampling period T_s. The resulting voltage-time system with the normalised duty cycles t_j (period of time in which S_j is set to 1 within a sampling period) respect to T_s and the reference voltages v_m^* is the following:

$$\begin{aligned} v_\alpha^* &= (2/3)(t_a + C_\delta t_d - S_\delta t_b - C_\delta t_e - S_\delta t_c) \\ v_\beta^* &= (2/3)(S_\delta t_d + C_\delta t_b + S_\delta t_e - C_\delta t_c - t_f) \\ v_x^* &= (2/3)(t_a - C_\delta t_d - S_\delta t_b + C_\delta t_e - S_\delta t_c) \\ v_y^* &= (2/3)(S_\delta t_d - C_\delta t_b + S_\delta t_e + C_\delta t_c - t_f) \end{aligned} \quad (5)$$

The modulation law can be arranged by means of elemental operations in order to achieve an equivalent linear equation system (the solution is not affected). Two decoupled and simpler modulation laws are obtained by combining appropriately the equations of (5). The simplified result is summarised as follows:

$$\begin{aligned} (v_\alpha^* + v_x^*) &= (4/3)(t_a - S_\delta t_b - S_\delta t_c) \\ (v_\beta^* - v_y^*) &= (4/3)(C_\delta t_b - C_\delta t_c) \end{aligned} \quad (6)$$

$$\begin{aligned} -(v_\beta^* + v_y^*) &= (4/3)(t_f - S_\delta t_d - S_\delta t_e) \\ (v_\alpha^* - v_x^*) &= (4/3)(C_\delta t_d - C_\delta t_e) \end{aligned} \quad (7)$$

The systems of (6) and (7) are two independent three-phase modulation spaces according to the approach of [22]. Consequently, they can be implemented with the fast algorithm developed in the cited work. A description of the method is included in Appendix A. In this scheme, the left-side of the identities are the stationary reference voltages. Then, the modulators should be commanded with the following auxiliary target signals:

$$\begin{aligned} v_{d1}^* &= v_\alpha^* + v_x^* \\ v_{q1}^* &= v_\beta^* - v_y^* \\ v_{d2}^* &= -(v_\beta^* + v_y^*) \\ v_{q2}^* &= v_\alpha^* - v_x^* \end{aligned} \qquad (8)$$

Additionally, notice that the model of the second modulator, see the first equation of (7), indicates that the first output of this block controls the S_f switching signal, whereas the remaining ought to be linked to S_d and S_e, respectively.

The block diagram of the proposed strategy is depicted in Figure 2. The two three-phase modulators are commanded by the references voltages defined in (8) along with the zero-sequence control signals λ_1 and λ_2. These last are set to 0.50 to provide the SVM operation, see Table A1, as an introduction of the method. Thus, the first modulator, MOD1, is configured with the references v_{d1}^* and v_{q1}^* to compute the duty cycles of the legs controlled by the S_a, S_b and S_c switching signals of the six-phase VSC. The MOD2 is commanded with v_{d2}^* and v_{q2}^*, while the transposition of the switching signal previously described is implemented. The three-phase techniques attain the duty times with Algorithm A1, detailed in the appendix. This information is employed by a six-channel digital PWM peripheral, which works with up/down counters and comparators to generate the gating signals. Theses activate/deactivate the power switches to synthesise the multifrequency output voltage.

Let us consider the following example to illustrate the operation of the proposed technique. The references voltages are $v_\alpha^* = 0.3653$, $v_\beta^* = 0.9309$, $v_x^* = 0.0956$, and $v_y^* = -0.0295$. Hence, the auxiliary references voltages defined by (8) are:

$$\begin{aligned} v_{d1}^* &= v_\alpha^* + v_x^* = 0.3653 + 0.0956 = 0.4609 \\ v_{q1}^* &= v_\beta^* - v_y^* = 0.9309 + 0.0295 = 0.9604 \\ v_{d2}^* &= -(v_\beta^* + v_y^*) = -(0.9309 - 0.0295) = -0.9014 \\ v_{q2}^* &= v_\alpha^* - v_x^* = 0.3653 - 0.0956 = 0.2697 \end{aligned} \qquad (9)$$

Then, the operation of the three-phase modulators with $\lambda_1 = \lambda_2 = 0.50$ is summarised as follows:

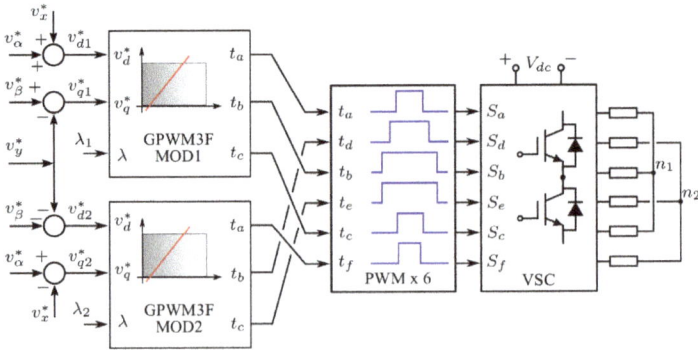

Figure 2. Generalised pulse width modulation (PWM) for asymmetrical six-phase VSCs.

MOD1:
$\tau_d = C_\delta \cdot |v_{q1}| = 0.8317$
$u^* = C_\delta^2 \cdot v_{d1} + S_\delta \cdot \tau_d = 0.7615$
$0 < u^* \le \tau_d$, then:
$\tau_{11} = u^* = 0.7615$
$a = 1 - \tau_d = 0.1683$
$t_a = \tau_{11} + a \cdot \lambda_1 = 0.8457$
$t_b = t_a - C_\delta \cdot (C_\delta \cdot v_{d1}^* - S_\delta \cdot v_{q1}^*) = 0.9159$
$t_c = t_b - C_\delta \cdot v_{q1}^* = 0.0841$

MOD2:
$\tau_d = C_\delta \cdot |v_{q2}| = 0.2336$
$u^* = C_\delta^2 \cdot v_{d2} + S_\delta \cdot \tau_d = -0.5593$
$-C_\delta \le u^* \le 0$, then:
$\tau_{11} = 0$
$a = 1 + u^* - \tau_d = 0.2072$
$t_f = \tau_{11} + a \cdot \lambda_2 = 0.1036$
$t_d = t_f - C_\delta \cdot (C_\delta \cdot v_{d2}^* - S_\delta \cdot v_{q2}^*) = 0.8964$
$t_e = t_d - C_\delta \cdot v_{q2}^* = 0.6628$

Notice that the operation of the proposal is simple and controls all the degree of freedom of the multiphase VSC. The output (duty cycles) is suitable for the implementation of the algorithm with PWM peripheral of the digital controllers. All these features are promising for the promotion of the multiphase technology in the industrial sector.

4. Experimental Validation and Discussions

The objective of this section is to provide the experimental proofs of the proper operation of the developed method. The architecture is composed by two three-phase generalised PWM blocks, whose performance was assessed only with single frequency tests. In this work, this scheme is commanded by reference voltages composed of a mix between two independent magnitudes and frequencies. Then, the proper generation of the output voltage in the time and frequency domains ought to be verified to demonstrated its viability for multiphase applications.

The developed modulator is assessed in the experimental test rig indicated in Figure 3. This is a six-phase VSC built with the FGH80N60FDTU IGBT. The load employed for the tests is a resistor-inductor, which is the most common behaviour in multiphase applications. The parameters of the experimental setup are detailed in Table 1. The dc-bus voltage is attained with a three-phase diode bridge rectifier power-supplied by a variac to control the output voltage, which is filtered by two capacitor in series located in the dc-side. The voltage V_{dc} is regulated to 100 (V) in order to avoid overcurrents during the tests with the maximum dc-bus utilisation. The frequency index ($m_f = f_s / f_1$) defines the sampling frequency $f_s = 1.5$ (kHz) and period $T_s = 1/f_s$, while the fundamental frequency is 50 (Hz). The controller is the experimenter kit of Texas Instruments based on the digital signal processor (DSP) TMS320F28335, while the commutations and protections are in charge of the Nexys 3 Spartan-6 FPGA Trainer Board. The algorithm is programmed in the DSP that send the six PWM signals to the FPGA. This last generates the twelve switching states with a dead-time of 2 (μs) and stops the operation under the presence of a trip-zone alarm. An optical link between the digital and power stages is implemented to reduce the electromagnetic interference impact. Two three-phase VSCs interconnected by the dc-bus in a modular design supply the load. The architecture of the prototype is flexible for future expansion to study drives with a higher number of phases or implement modulation techniques with unconventional switching strategies. The Keysight DSOX3024T oscilloscope is employed to acquire the measurements along with the Fluke i3000s Flex-24 ac current clamp and the Elditest differential voltage probe GE 8100.

Table 1. Parameters for the experimental tests.

Parameter	Unit	Value
Resistance, R	(Ω)	10
Inductance, L	(H)	10 m
Dc-bus Voltage, V_{dc}	(V)	100
Fundamental frequency, f_1	(Hz)	50
Frequency index, m_f		30

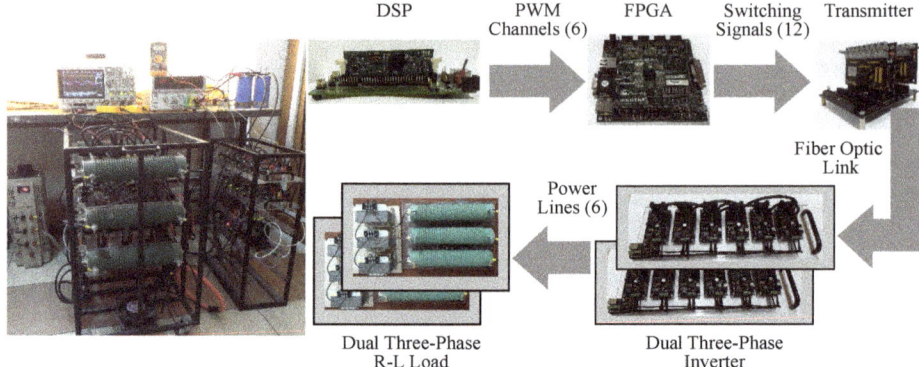

Figure 3. Experimental setup for the validation process.

The magnitude of the reference voltages are controlled with the fundamental and harmonic modulation indexes; m_1 and m_2, respectively. The second voltage is set with the fifth-order harmonic frequency (250 Hz) in order to inject this component in the x-y subspace. Diverse combinations are considered to provide an assessment in a wide operation range. The experimental tests for the evaluation of the time and frequency response of the scheme are described as follows.

First, the single frequency test with the maximum modulation index [10], $m_1 = 1.1547$ and $m_2 = 0$, is depicted in Figure 4. The traditional switching pattern and sinusoidal current can be appreciated in the time response. The voltage spectrum attained with the discrete Fourier transform from 0 to 5 (kHz) and 500 (Hz) per division is also included in capture of the oscilloscope. This shows a clean spectrum in the low-order frequencies interval. The harmonics components are only the effect of the modulation process in the sidebands around the carrier frequency (m_f-order harmonic component).

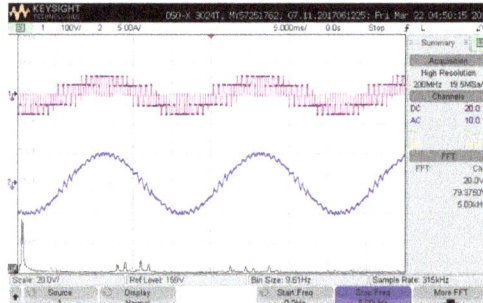

Figure 4. Voltage (top) and current (middle) waveforms of the phase a along with the V_a spectrum (bottom) for the reference voltage $m_1 = 1.15$ and $m_2 = 0$.

The proposed algorithm is capable of operating in the multifrequency mode. Then, the reference voltage of the modulator is set with a fundamental frequency of $m_1 = 0.92$ and a fifth-order harmonic of $m_2 = 0.23$. This combination provides the maximum dc-bus utilisation in the linear modulation region since $m_1 + m_2 = 1.1547$ [23]. The behaviour achieved is presented in Figure 5. The phase voltage has a slightly different switching pattern with respect to the previous case to synthesise the target voltage of the x-y plane. The current waveform denotes the frequency mixing with non sinusoidal behaviour. On the other hand, the fast Fourier transform (FFT) proofs the low harmonic energy retained in the synthesised voltage. Furthermore, the sidebands around $2f_s$ are the more notable components, whereas for the single frequency this happens in the neighbourhood of the carrier frequency.

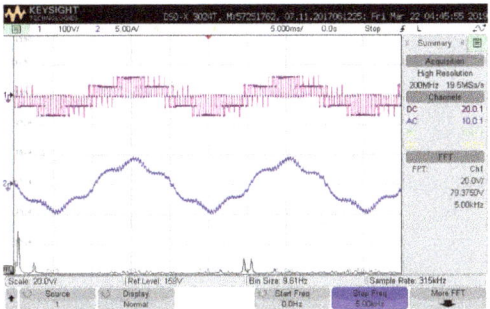

Figure 5. Voltage (top) and current (middle) waveforms of the phase *a* along with the voltage spectrum (bottom) for the reference voltage $m_1 = 0.92$ and $m_2 = 0.23$.

The most extreme combination for multiphase applications is $m_1 = m_2$. The response achieved for this case when both magnitudes are 0.57 to reach the maximum dc-bus utilisation is indicated in Figure 6. The voltage switching pattern is completely different respect to the conventional case, attaining a more dynamic waveform. The current behaviour in the time-domain also evidences the multifrequency operation. The frequency response again reproduces the previous features, generating undesired harmonics components only in the sidebands. The harmonic energy is practically distributed in equal proportions between f_s and $2f_s$ in comparison with the last spectrum.

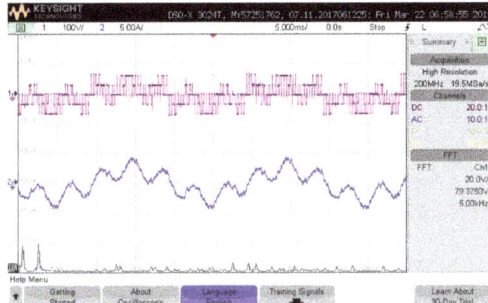

Figure 6. Time response of the voltage (upper graph) and current (middle waveform) of the phase *a* along with the voltage spectrum (bottom) for the reference voltage $m_1 = m_2 = 0.57$.

The control strategy in the synchronous frame to mitigate the dead-time effects of the real multiphase drive regulates the current of the *x-y* plane [18]. Therefore, the voltage that should be synthesised by the modulator in this use might be composed of a combination of the harmonic components that engage this subspace. An experimental test with simultaneous injection of the 5th- and 7th-order harmonics is carried out as a proof of concept of the proposal for the considered application, see Figure 7. Thus, the fundamental frequency is regulated with $m_1 = 0.90$ and the reference voltage in the *x-y* plane is the result of the mix between the two harmonics configured with magnitudes of 0.15 and 0.10, respectively. The response of the developed algorithm is appropriate for this case showing the expected waveforms, while the generation of fundamental and the two target harmonics components in agreement with their respective set-points are confirmed in the spectrum. Notice that the impact of the modulation process is low, and only the amplitude of the sidebands locate around 3 (kHz), $2f_s$, are perceptible.

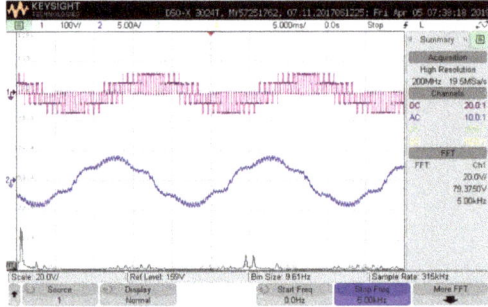

Figure 7. Voltage (upper graph) and current (middle waveform) of phase *a* along with the synthesised voltage spectrum (bottom) with a fundamental frequency magnitude of 0.90, while the 5th- and 7th-order harmonics are set to 0.15 and 0.10, respectively.

The magnitudes of the synthesised voltages in the previous tests are in agreement with the command values according to the obtained spectrum. Moreover, the time-domain response has the expected behaviour. Then, an analysis of the voltage and current distortion is the next step in this evaluation process. As two frequency components are injected, a compound total harmonic distortion (CTHD) is proposed to assess the method. This performance parameter is calculated as follows:

$$\text{CTHD}_f = \frac{\sqrt{\sum_{h \neq 1,5} f_h^2}}{\sqrt{f_1^2 + f_5^2}} \tag{10}$$

where f can be either the current or the voltage and the subscript h designates the h-order harmonic component. Then, this metric provides the proportion of the total amount of undesired harmonics achieved with respect to the injected magnitudes.

The ratio $r = m_2/m_1$ is defined to conduct the experimental results in this part. Then, r is adjusted with the following values 0.00, 0.10, 0.20, 0.40 and 1.00 to assess a wide range of operation points with the developed method. For each of these ratios, the exploration is performed by fixing r and varying the magnitude of m_1 from 0.30 up to the limit of the dc-bus utilisation, including also this last. The test rig is configured with these combinations of modulation indexes and the data waveform attained with the oscilloscope is stored to be processed. In this manner, the CTHD_v and CTHD_i are computed by applying (10) with the measurements.

The behaviour of the CTHD_v achieved with the technique is illustrated in Figure 8. The single frequency tests provide the conventional total harmonic distortion [24], with decreasing values in all the intervals. The response of the scheme with ratios higher than zero shows similar characteristics. For $r < 0.50$ the CTHD_v is practically identical to the first case, whereas for higher ratios the result is noticeably lower, see $r = 1.00$. The reason is the resulted sidebands of the spectrum. Notice that in the previous results of Figure 6, $m_1 = m_2$ and $r = 1$, the harmonic's energy is distributed around f_s and $2f_s$, while for the other combinations, Figures 4 and 5, this is concentrated only in one of these frequencies. This effect generates a larger voltage harmonics that produce higher CTHD_v.

The measured currents are processed with (10) and the results are summarised in Figure 9. In all the cases, CTHD_i decreases with respect to m_1. The notorious characteristic is the higher current distortion achieved for $r = 1.00$ in spite of reporting the lower CTHD_v values in the previous test. Again, the frequency response of the voltage ought to be analysed to find the reasons. The voltage spectrum generated with $m_1 = m_2$ has smaller harmonic components, but they are distributed around f_s and $2f_s$. In the other cases ($r < 0.50$) the concentration is in the neighbourhood of $2f_s$. Since the impedance is proportional to the frequency, the sidebands of $r = 1.00$ present lower opposition and generate higher harmonic currents, which impact the reckoning of CTHD_i.

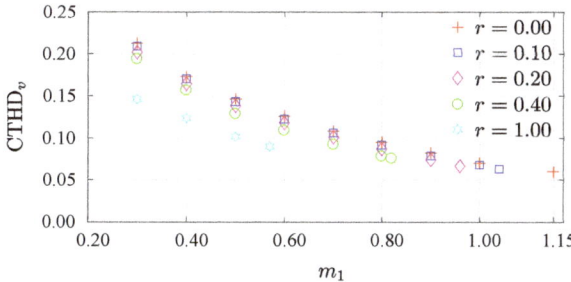

Figure 8. compound total harmonic distortion (CTHD$_v$) for different operation points within the linear modulation region.

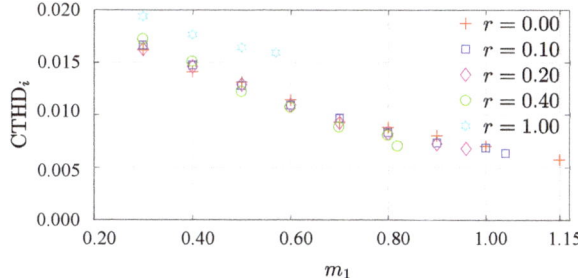

Figure 9. CTHD$_i$ for different operation points within the linear modulation region.

The time and spectrum behaviour of the multifrequency modulation strategy with continuous switching mode (two commutations per sampling period) was evaluated with more than 40 experimental tests. The degrees of freedom were configured to operate as the SVM for the sake of simplicity in the introduction of the technique. The undesired frequencies are present in the sidebands, caused by the mix between the carrier frequency, the fundamental and the harmonics injected in the x-y plane (modulation process). In this manner, the behaviour of the spectrum is available to design power filters or to mitigate possible sources of resonance, achieving robust electric drives. This section demonstrates the viability and great potential of the algorithm by following a rigorous procedure.

5. Conclusions

A multifrequency modulation technique for dual three-phase voltage source converters has been developed in this paper. The scheme is based on a generalised solution of the voltage-time law and implemented with a simple algorithm that computes the duty cycles of the power switches immediately. This output can be integrated straightforwardly with the pulse width modulation peripherals of the digital controllers. Validation tests to analyse the time and frequency response were conducted by injecting fundamental and fifth-order frequency components. The proposal demonstrates good magnitude tracking with a small number of undesired harmonic components, which are generated due to the modulation process. The current distortion attained (associated with the current ripple) is also low and was measured with a proposed figure of merit. The promising features of the method for multiphase power electronics applications can be proved in this manner.

Author Contributions: Conceptualization, J.A.R. and J.P.; methodology, J.A.R., M.R. and R.G.; software, J.A.R., S.T. and M.R.; validation, J.P. and J.A.R.; formal analysis, R.G., M.R. and J.P.; investigation, J.A.R., J.P. and M.R.; resources, M.R., R.G. and S.T.; data curation, M.R. and R.G.; writing—original draft preparation, J.A.R., M.R. and S.T.; writing—review and editing, R.G., M.R., S.T. and J.P.; visualization, J.A.R., M.R. and S.T.; supervision, J.A.R. and R.G.; project administration, R.G., J.A.R. and M.R.; funding acquisition, S.T., R.G., M.R. and J.A.R.

Funding: This research was funded in part by Paraguayan Program for the Development of Science and Technology (PROCIENCIA) through the R&D project component with reference 14-INV-097, in part by the Chilean Fund of Scientific and Technological Development (FONDECYT) under the grants provided in the Postdoctoral 3170014 and the Regular 1160690 projects.

Conflicts of Interest: The authors declare no conflict of interest.

Appendix A. Generalised Modulation Scheme for Three-Phase VSC

A three-phase modulation space described by the switching signals, and represented by using the constants previously defined in this paper, is given by [22]:

$$v_d = (4/3)(S_a - S_\delta S_b - S_\delta S_c)$$
$$v_q = (4/3)(C_\delta S_b - C_\delta S_c)$$
(A1)

where v_d and v_q are the normalised voltages space vector components in the stationary reference frame generated by the switching vector $[S_a\ S_b\ S_c]$. Then, the modulation law of this system described with the normalised duty cycles $t_j = T_j/T_s$ of the switching signals along with the reference voltages v_d^* and v_q^* is presented as follows:

$$v_d^* = (4/3)(t_a - S_\delta t_b - S_\delta t_c)$$
$$v_q^* = (4/3)(C_\delta t_b - C_\delta t_c)$$
(A2)

The system of linear equation attained in (A2) could have an infinite number of solutions within the domain of $t_j \in [0,1]$ and the linear modulation region. The geometric interpretation is the intersection line of two three-dimensional planes, which can be determined by an appropriate transformation in a two-dimensional auxiliary space. A generalised analytical solution was proposed in [22] by adding the $\lambda \in [0,1]$ degree of freedom, that is the zero-sequence voltage of the electric modelling. Thus, the duty times can be computed by applying Algorithm A1.

Algorithm A1: Generalised PWM for three-phase inverters

    Inputs: v_d^*, v_q^*, λ
    Constants and variables: $C_\delta, S_\delta, \tau_d, u^*, \tau_{11}, a$
    Outputs: t_a, t_b, t_c
    // Calculate the algorithm constants
    $\tau_d = C_\delta \cdot |v_q^*|$
    $u^* = C_\delta^2 \cdot v_d^* + S_\delta \cdot \tau_d$
    // Calculate the values of τ_{11} and a
    if $(-C_\delta \leq u^* \leq 0)$ **then**
        $\tau_{11} = 0$
        $a = u^* + 1 - \tau_d$
    else if $(0 < u^* \leq \tau_d)$ **then**
        $\tau_{11} = u^*$
        $a = 1 - \tau_d$
    else if $(\tau_d < u^* \leq 1)$ **then**
        $\tau_{11} = u^*$
        $a = 1 - u^*$
    end if
    //Calculate the duty cycles
    $t_a = \tau_{11} + a \cdot \lambda$
    $t_b = t_a - C_\delta \cdot (C_\delta \cdot v_d^* - S_\delta \cdot v_q^*)$
    $t_c = t_b - C_\delta \cdot v_q^*$

Table A1. Operation modes of the generalised pulse width modulation (PWM)

λ	PWM Technique
0	PWM-Min
1/2	SVM
1	PWM-Max

The inputs of the presented algorithm are the normalised reference voltages and the zero-sequence control variable. Then, the auxiliary constants τ_d and u^* are calculated with simple operators. These are processed by a conditional routine with three possible cases to attain the auxiliary values of τ_{11} and a, respectively. These are used to compute t_a, whereas this value is applied into (A2) to solve the linear equation system and obtain the remaining duty cycles. Notice that the architecture is simple and only employs low computational cost operations.

The degree of freedom λ has been identified as the proportion between the dwell time of the vector [1 1 1] and the total application time of the zero voltage vector, which is also distributed with the vector [0 0 0]. Hence, the generalised modulator can be configured to operate like the well-known techniques [25] by selecting the values of λ presented in Table A1.

References

1. Duran, M.J.; Levi, E.; Barrero, F. Multiphase Electric Drives: Introduction. In *Wiley Encyclopedia of Electrical and Electronics Engineering*; John Wiley & Sons Ltd.: Hoboken, NJ, USA, 2017; pp. 1–26.
2. Levi, E. Advances in Converter Control and Innovative Exploitation of Additional Degrees of Freedom for Multiphase Machines. *IEEE Trans. Ind. Electron.* **2016**, *63*, 433–448. [CrossRef]
3. Barrero, F.; Duran, M. Recent Advances in the Design, Modeling, and Control of Multiphase Machines—Part I. *IEEE Trans. Ind. Electron.* **2016**, *63*, 449–458. [CrossRef]
4. Duran, M.; Barrero, F. Recent Advances in the Design, Modeling, and Control of Multiphase Machines—Part II. *IEEE Trans. Ind. Electron.* **2016**, *63*, 459–468. [CrossRef]
5. Liu, Z.; Li, Y.; Zheng, Z. A review of drive techniques for multiphase machines. *CES Trans. Electr. Mach. Syst.* **2018**, *2*, 243–251. [CrossRef]
6. Wang, Z.; Wang, Y.; Chen, J.; Hu, Y. Decoupled Vector Space Decomposition Based Space Vector Modulation for Dual Three-Phase Three-Level Motor Drives. *IEEE Trans. Power Electron.* **2018**, *33*, 10683–10697. [CrossRef]
7. Gutierrez-Reina, D.; Barrero, F.; Riveros, J.; Gonzalez-Prieto, I.; Toral, S.L.; Duran, M.J. Interest and Applicability of Meta-Heuristic Algorithms in the Electrical Parameter Identification of Multiphase Machines. *Energies* **2019**, *12*. [CrossRef]
8. Hu, Y.; Zhu, Z.Q.; Odavic, M. Torque Capability Enhancement of Dual Three-Phase PMSM Drive with Fifth and Seventh Current Harmonics Injection. *IEEE Trans. Ind. Appl.* **2017**, *53*, 4526–4535. [CrossRef]
9. Che, H.S.; Duran, M.J.; Levi, E.; Jones, M.; Hew, W.; Rahim, N.A. Postfault Operation of an Asymmetrical Six-Phase Induction Machine With Single and Two Isolated Neutral Points. *IEEE Trans. Power Electron.* **2014**, *29*, 5406–5416. [CrossRef]
10. Levi, E.; Dujic, D.; Jones, M.; Grandi, G. Analytical Determination of DC-Bus Utilization Limits in Multiphase VSI Supplied AC Drives. *IEEE Trans. Energy Convers.* **2008**, *23*, 433–443. [CrossRef]
11. Gonzalez-Prieto, I.; Duran, M.J.; Aciego, J.J.; Martin, C.; Barrero, F. Model Predictive Control of Six-Phase Induction Motor Drives Using Virtual Voltage Vectors. *IEEE Trans. Ind. Electron.* **2018**, *65*, 27–37. [CrossRef]
12. Zhao, Y.; Lipo, T.A. Space vector PWM control of dual three-phase induction machine using vector space decomposition. *IEEE Trans. Ind. Appl.* **1995**, *31*, 1100–1109. [CrossRef]
13. Hadiouche, D.; Baghli, L.; Rezzoug, A. Space-vector PWM techniques for dual three-phase AC machine: analysis, performance evaluation, and DSP implementation. *IEEE Trans. Ind. Appl.* **2006**, *42*, 1112–1122. [CrossRef]
14. Marouani, K.; Baghli, L.; Hadiouche, D.; Kheloui, A.; Rezzoug, A. A New PWM Strategy Based on a 24-Sector Vector Space Decomposition for a Six-Phase VSI-Fed Dual Stator Induction Motor. *IEEE Trans. Ind. Electron.* **2008**, *55*, 1910–1920. [CrossRef]

15. Yazdani, D.; Khajehoddin, S.A.; Bakhshai, A.; Joos, G. Full Utilization of the Inverter in Split-Phase Drives by Means of a Dual Three-Phase Space Vector Classification Algorithm. *IEEE Trans. Indu. Electron.* **2009**, *56*, 120–129. [CrossRef]
16. Zhou, C.; Yang, G.; Su, J. PWM Strategy With Minimum Harmonic Distortion for Dual Three-Phase Permanent-Magnet Synchronous Motor Drives Operating in the Overmodulation Region. *IEEE Trans. Power Electron.* **2016**. *31*, 1367–1380. [CrossRef]
17. Prieto, J.; Levi, E.; Barrero, F.; Toral, S. Output current ripple analysis for asymmetrical six-phase drives using double zero-sequence injection PWM. In Proceedings of the 37th Annual Conference of the IEEE Industrial Electronics Society, Melbourne, VIC, Australia, 7–10 November 2011; pp. 3692–3697. [CrossRef]
18. Jones, M.; Vukosavic, S.N.; Dujic, D.; Levi, E. A Synchronous Current Control Scheme for Multiphase Induction Motor Drives. *IEEE Trans. Energy Convers.* **2009**, *24*, 860–868. [CrossRef]
19. Prieto, J.; Riveros, J.A.; Bogado, B. Multifrequency output voltage generation for asymmetrical dual three-phase drives. In Proceedings of the IEEE International Electric Machines and Drives Conference (IEMDC), Miami, FL, USA, 21–24 May 2017; pp. 1–6. [CrossRef]
20. Riveros, J.A.; Prieto, J.; Guzman, H. Multifrequency space vector pulse width modulation for asymmetrical six-phase drives. In Proceedings of the IEEE 26th International Symposium on Industrial Electronics (ISIE), Edinburgh, UK, 19–21 June 2017; pp. 781–786. [CrossRef]
21. Dujic, D.; Grandi, G.; Jones, M.; Levi, E. A Space Vector PWM Scheme for Multifrequency Output Voltage Generation With Multiphase Voltage-Source Inverters. *IEEE Trans. Ind. Electron.* **2008**, *55*, 1943–1955. [CrossRef]
22. Riveros, J.A.; Prieto, J.; Rivera, M. Generalised pulse width modulation algorithm for three-phase voltage source converters. In Proceedings of the IEEE Southern Power Electronics Conference (SPEC), Puerto Varas, Chile, 4–7 December 2017; pp. 1–6.
23. Dujic, D.; Jones, M.; Levi, E.; Lopez, O. DC-bus utilisation in series-connected multi-phase machines supplied from a VSI with a composite phase number. In Proceedings of the XIX International Conference on Electrical Machines—ICEM 2010, Rome, Italy, 6–8 September 2010, pp. 1–6.
24. Engku Ariff, E.; Dordevic, O.; Jones, M. A Space Vector PWM Technique for a Three-Level Symmetrical Six-Phase Drive. *IEEE Trans. Ind. Electron.* **2017**, *64*, 8396–8405. [CrossRef]
25. Holmes, D.G.; Lipo, T.A. *Pulse Width Modulation for Power Converters: Principles and Practice*; Wiley-IEEE Press: Piscataway, NJ, USA, 2003; Chapter 6, pp. 259–336.

© 2019 by the authors. Licensee MDPI, Basel, Switzerland. This article is an open access article distributed under the terms and conditions of the Creative Commons Attribution (CC BY) license (http://creativecommons.org/licenses/by/4.0/).

Article

Current Control of a Six-Phase Induction Machine Drive Based on Discrete-Time Sliding Mode with Time Delay Estimation

Yassine Kali [1,*], **Magno Ayala** [2], **Jorge Rodas** [2], **Maarouf Saad** [1], **Jesus Doval-Gandoy** [3], **Raul Gregor** [2] **and Khalid Benjelloun** [4]

1. Power Electronics and Industrial Control Research Group (GRÉPCI), École de Technologie Supérieure, Montreal H3C 1K3, QC, Canada; maarouf.saad@etsmtl.ca
2. Laboratory of Power and Control Systems (LSPyC), Facultad de Ingeniería, Universidad Nacional de Asunción, Luque 2060, Paraguay; mayala@ing.una.py (M.A.); jrodas@ing.una.py (J.R.); rgregor@ing.una.py (R.G.)
3. Applied Power Electronics Technology Research Group (APET), Universidad de Vigo, 36310 Vigo, Spain; jdoval@uvigo.es
4. Electrical Engineering Department, Ecole Mohammadia d'Ingénieurs, University of Mohammed V, Rabat 765, Morocco; bkhalid@emi.ac.ma
* Correspondence: y.kali88@gmail.com; Tel.: +1-514-443-8118

Received: 22 November 2018; Accepted: 29 December 2018; Published: 5 January 2019

Abstract: This paper proposes a robust nonlinear current controller that deals with the problem of the stator current control of a six-phase induction motor drive. The current control is performed by using a state-space representation of the system, explicitly considering the unmeasurable states, uncertainties and external disturbances. To estimate these latter effectively, a time delay estimation technique is used. The proposed control architecture consists of inner and outer loops. The inner current control loop is based on a robust discrete-time sliding mode controller combined with a time delay estimation method. As said before, the objective of the time delay estimation is to reconstruct the unmeasurable states and uncertainties, while the sliding mode aims is to suppress the estimation error, to ensure robustness and finite-time convergence of the stator currents to their desired references. The outer loop is based on a proportional-integral controller to control the speed. The stability of the current closed-loop system is proven by establishing sufficient conditions on the switching gains. Experimental work has been conducted to verify the performance and the effectiveness of the proposed robust control scheme for a six-phase induction motor drive. The results obtained have shown that the proposed method allows good performances in terms of current tracking, in their corresponding planes.

Keywords: multiphase induction machine; time delay estimation; sliding mode control; field-oriented control; current control

1. Introduction

Multiphase drives have received significant interest from the power electronics, control, machines and drives communities due to their good features in comparison with traditional three-phase drives. The features include lower torque ripple, lower current/power per phase and fault-tolerant capabilities without adding extra hardware [1–3]. Currently, multiphase drives are extensively used in several applications where high power is required such as ships, wind energy generation systems and electric vehicles [3,4]. In the literature, most of the developed and published control techniques for multiphase Induction Machine (IM) drives are an extension of the ones designed for the three-phase machines such as Proportional-Resonant (PR) [5], Proportional-Integral (PI) Pulse-Width Modulation (PWM) [6],

Direct Torque Control (DTC) [7], Predictive Torque Control (PTC) [8], sensorless [9,10] and Model Predictive Control (MPC) [11,12], among others. Recently, the above-mentioned controllers have been extended for multiphase machines under fault situations [13–16]. However, few published papers have considered robust nonlinear controllers and intelligent techniques such as backstepping [17,18], Sliding Mode Control (SMC) [19–21], fuzzy logic [22] and others.

Among the above-mentioned nonlinear controllers, SMC is one of the most widely used and has received particular attention from the automation community due to its three highly-valued properties, namely robustness against matched uncertainties, simplicity of design and finite-time convergence [23,24]. This method forces the system states to reach in finite time the user-selected sliding surface (switching surface) even in the presence of the matched uncertainties using discontinuous inputs [24]. To ensure high performances, the switching gains should be chosen as large as possible to reject the effect of the bounded uncertainties. However, this choice causes the major drawback of SMC, well-known under the name of the chattering phenomenon [25,26]. The latter has an unpleasant impact on system actuators. It can lead to deterioration of the controlled system and/or instability. Once this problem has been identified, many works that tried to solve it were published, and among them, we cite the following:

- The substitution of the discontinuous signum function by linear ones [27]. This method is the well-known SMC based on a boundary layer. This proposition allows the reduction of the chattering phenomenon. However, the finite-time convergence feature is no longer guaranteed. The latter is very desirable when critical convergence time is required.
- Observer-based SMC [28,29]. The issue of designing a robust nonlinear controller in this technique is reduced to the issue of designing a robust nonlinear observer. In other words, if the matched uncertainties are not accurately estimated, the performances obtained will not be satisfactory.
- Higher Order Sliding Mode (HOSM) [30–32]. The idea consists of making the switching control term act on the control input derivative, which makes the control input fed into the system continuous. This method gives better performances since it allows higher precision and reduces the chattering phenomenon. However, this approach requires some information, as the first time derivative of the selected sliding surface is not always available for measurements, making the implementation difficult.

Recently, an interesting method that consists of combining SMC with the Time Delay Estimation (TDE) method for uncertain nonlinear systems [33,34] has been developed. The proposed method has been successfully tested on a redundant robot manipulator. The basic idea is to estimate the matched uncertainties that are assumed to be Lipschitz using delayed states and input information. Then, the estimated terms are added to the equivalent controller in order to allow a small choice of the switching gains of the discontinuous controller.

Nevertheless, real-time implementation is generally performed through discrete systems [35]. For this reason, the development of the controller should be done in discrete-time. Consequently, it is suitable for use with a discrete-time model of the six-phase IM during the design procedure since after discretization, the inherent properties of the sliding mode approach can no longer be maintained.

In summary, the aim of this paper is to develop a robust Discrete-time SMC (DSMC) combined with the TDE method for the inner current control loop of an Indirect Rotor Field-Oriented Control (IRFOC) of a six-phase IM drive. The developed controller works for all multiphase machines in several applications as more electric aircraft, ship propulsion, battery-powered electric vehicles, electric traction and hybrid electric vehicles. Experimental validation is presented to show the effectiveness of the current controller in transient and steady-state conditions. The rest of the paper is organized as follows. The mathematical discrete-time model of the considered system is presented in Section 2, while the proposed controller design and detailed stability analysis are explained in Section 3. Experimental results are presented in Section 4. Finally, Section 5 draws some conclusions.

2. Six-Phase IM and VSI Model

The considered system shown in Figure 1 consists of the asymmetrical six-phase IM fed by two two-Level (2L) Voltage Source Inverter (VSI). After using the Vector Space Decomposition (VSD) approach, the decoupling transformation **T** gives the $\alpha - \beta$ subspace, which is related to the flux/torque producing components and the loss-producing $x - y$ subspace and a zero-sequence subspace. Then, by using an amplitude-invariant transformation matrix, **T** is defined as follows:

$$\mathbf{T} = \frac{1}{3} \begin{bmatrix} 1 & \frac{\sqrt{3}}{2} & -\frac{1}{2} & -\frac{\sqrt{3}}{2} & -\frac{1}{2} & 0 \\ 0 & \frac{1}{2} & \frac{\sqrt{3}}{2} & \frac{1}{2} & -\frac{\sqrt{3}}{2} & -1 \\ 1 & -\frac{\sqrt{3}}{2} & -\frac{1}{2} & \frac{\sqrt{3}}{2} & -\frac{1}{2} & 0 \\ 0 & \frac{1}{2} & -\frac{\sqrt{3}}{2} & \frac{1}{2} & \frac{\sqrt{3}}{2} & -1 \\ 1 & 0 & 1 & 0 & 1 & 0 \\ 0 & 1 & 0 & 1 & 0 & 1 \end{bmatrix}. \tag{1}$$

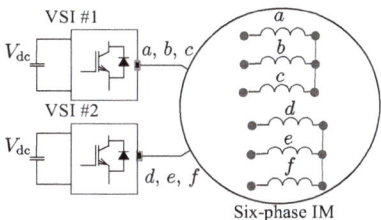

Figure 1. Scheme of the six-phase induction machine drive.

The discrete-time model of the system in state-space representation is represented by the following equations [36]:

$$\mathbf{X}(k+1) = \mathbf{A}\,\mathbf{X}(k) + \mathbf{B}\,\mathbf{u}(k) + \mathbf{n}(k) \tag{2}$$
$$\mathbf{Y}(k) = \mathbf{C}\,\mathbf{X}(k) \tag{3}$$

In the equations above, the stator and rotor currents are the state vector:

$$\mathbf{X}(k) = \left[i_{s\alpha}(k), i_{s\beta}(k), i_{sx}(k), i_{sy}(k), i_{r\alpha}(k), i_{r\beta}(k)\right]^T \tag{4}$$

while the stator voltages represent the input vector:

$$\mathbf{u}(k) = \left[u_{s\alpha}(k), u_{s\beta}(k), u_{sx}(k), u_{sy}(k)\right]^T \tag{5}$$

and the stator currents the output vector:

$$\mathbf{Y}(k) = \left[i_{s\alpha}(k), i_{s\beta}(k), i_{sx}(k), i_{sy}(k)\right]^T \tag{6}$$

and $\mathbf{n}(k)$ is the (6×1) uncertain vector. The stator voltages have a discrete nature due to the VSI model, and the relationship between them is represented as:

$$V_{dc}\,\mathbf{T}\,\mathbf{M} = \left[u_{s\alpha}(k), u_{s\beta}(k), u_{sx}(k), u_{sy}(k)\right]^T \tag{7}$$

where V_{dc} is the DC-bus voltage, and the VSI model is:

$$\mathbf{M} = \frac{1}{3} \begin{bmatrix} 2 & 0 & -1 & 0 & -1 & 0 \\ 0 & 2 & 0 & -1 & 0 & -1 \\ -1 & 0 & 2 & 0 & -1 & 0 \\ 0 & -1 & 0 & 2 & 0 & -1 \\ -1 & 0 & -1 & 0 & 2 & 0 \\ 0 & -1 & 0 & -1 & 0 & 2 \end{bmatrix} \mathbf{S}^T \quad (8)$$

where $\mathbf{S} = \begin{bmatrix} S_a, S_b, S_c, S_d, S_e, S_f \end{bmatrix}$ is the vector of the gating signals with $S_i \in \{0, 1\}$. Moreover, the matrices $\mathbf{A} \in R^{6 \times 6}$, $\mathbf{B} \in R^{6 \times 4}$ and $\mathbf{C} \in R^{4 \times 6}$ are defined by:

$$\mathbf{A} = \begin{bmatrix} a_{11} & a_{12} & 0 & 0 & a_{15} & a_{16} \\ a_{21} & a_{22} & 0 & 0 & a_{25} & a_{26} \\ 0 & 0 & a_{33} & 0 & 0 & 0 \\ 0 & 0 & 0 & a_{44} & 0 & 0 \\ a_{51} & a_{52} & 0 & 0 & a_{55} & a_{56} \\ a_{61} & a_{62} & 0 & 0 & a_{65} & a_{66} \end{bmatrix} \quad (9)$$

$$\mathbf{B} = \begin{bmatrix} b_1 & 0 & 0 & 0 \\ 0 & b_1 & 0 & 0 \\ 0 & 0 & b_2 & 0 \\ 0 & 0 & 0 & b_2 \\ b_3 & 0 & 0 & 0 \\ 0 & b_3 & 0 & 0 \end{bmatrix} \quad (10)$$

$$\mathbf{C} = \begin{bmatrix} 1 & 0 & 0 & 0 & 0 & 0 \\ 0 & 1 & 0 & 0 & 0 & 0 \\ 0 & 0 & 1 & 0 & 0 & 0 \\ 0 & 0 & 0 & 1 & 0 & 0 \end{bmatrix} \quad (11)$$

where:

$a_{11} = a_{22} = 1 - T_s c_2 R_s$ $a_{12} = -a_{21} = T_s c_4 L_m \omega_{r(k)}$ $a_{15} = a_{26} = T_s c_4 R_r$
$a_{16} = -a_{25} = T_s c_4 L_r \omega_{r(k)}$ $a_{33} = a_{44} = 1 - T_s c_3 R_s$ $a_{51} = a_{62} = T_s c_4 R_s$
$a_{52} = -a_{61} = -T_s c_5 L_m \omega_{r(k)}$ $a_{55} = a_{66} = 1 - T_s c_5 R_r$ $a_{56} = -a_{65} = -c_5 \omega_{r(k)} T_s L_r$
$b_1 = T_s c_2$ $b_2 = T_s c_3$ $b_3 = -T_s c_4$

with T_s the sampling time and c_1 to c_5 are defined as: $c_1 = L_s L_r - L_m^2$, $c_2 = \frac{L_r}{c_1}$, $c_3 = \frac{1}{L_{ls}}$, $c_4 = \frac{L_m}{c_1}$, $c_5 = \frac{L_s}{c_1}$. The electrical parameters of the systems are R_s, R_r, $L_r = L_{lr} + L_m$, $L_s = L_{ls} + L_m$, L_r and L_m. The rotor electrical speed ω_r is related to the load torque T_l and the generated torque T_e as follows:

$$J_m \dot{\omega}_r + B_m \omega_r = P(T_e - T_l) \quad (12)$$

$$\omega_r = P \omega_m \quad (13)$$

where J_m denotes the inertia coefficient, B_m denotes the friction coefficient, P denotes the number of pole pairs and the generated torque T_e is defined by:

$$T_e = 3 P \left(\psi_{s\alpha} i_{s\beta} - \psi_{s\beta} i_{s\alpha} \right) \quad (14)$$

where $\psi_{s\alpha}$ and $\psi_{s\beta}$ are the stator fluxes.

3. Controller Design and Stability Analysis

3.1. Outer Speed Control Loop

A two-degree PI controller with a saturation stage, introduced in [37], is used as the outer speed control loop, based on the IRFOC method. In this loop, the output of the PI regulator is used to get the dynamic current reference $i^*_{sq}(k)$. In addition, the slip frequency $\omega_{sl}(k)$ calculation is obtained from the current references $i^*_{sd}(k), i^*_{sq}(k)$ in the dynamic reference frame and the electrical parameters of the six-phase IM, as shown in Figure 2.

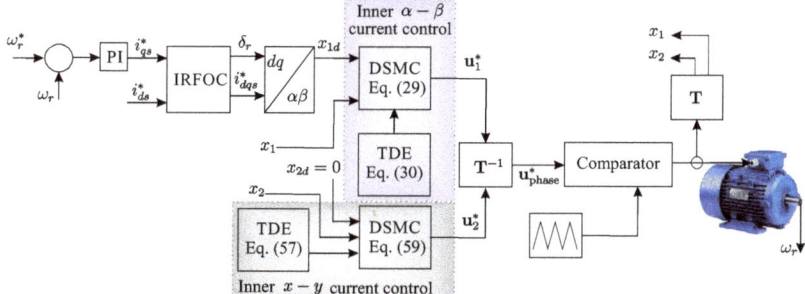

Figure 2. Block diagram of the closed-loop system based on IRFOC and the DSMC with TDE method.

3.2. Inner Current Control Loop

The inner loop aims to control the stator currents. For this purpose, the DSMC with TDE method will be derived to ensure the finite-time convergence of the stator currents in the $\alpha - \beta$ and the $x - y$ planes to their desired references with high accuracy even if some states are not measurable (i.e., rotor currents) and in the presence of uncertainties. First of all, let us decompose the discrete system described in (2) into three sub-systems as follows:

$$\mathbf{x}_1(k+1) = \mathbf{A}_1 \mathbf{x}_1(k) + \overline{\mathbf{A}}_1 \mathbf{x}_3(k) + \mathbf{B}_1 \mathbf{u}_1(k) + \eta_1(k) \quad (15)$$
$$\mathbf{x}_2(k+1) = \mathbf{A}_2 \mathbf{x}_2(k) + \mathbf{B}_2 \mathbf{u}_2(k) + \eta_2(k) \quad (16)$$
$$\mathbf{x}_3(k+1) = \mathbf{A}_3 \mathbf{x}_1(k) + \overline{\mathbf{A}}_3 \mathbf{x}_3(k) + \mathbf{B}_3 \mathbf{u}_1(k) + \eta_3(k) \quad (17)$$

where the stator and rotor current state vectors:

$$\mathbf{x}_1(k) = [i_{s\alpha}(k), i_{s\beta}(k)]^T \quad (18)$$
$$\mathbf{x}_2(k) = [i_{sx}(k), i_{sy}(k)]^T \quad (19)$$
$$\mathbf{x}_3(k) = [i_{r\alpha}(k), i_{r\beta}(k)]^T \quad (20)$$

while the stator voltages represent the input vectors:

$$\mathbf{u}_1(k) = [u_{s\alpha}(k), u_{s\beta}(k)]^T \quad (21)$$
$$\mathbf{u}_2(k) = [u_{sx}(k), u_{sy}(k)]^T \quad (22)$$

and $\eta_1(k) = [n_1(k), n_2(k)]^T, \eta_2(k) = [n_3(k), n_4(k)]^T$ and $\eta_3(k) = [n_5(k), n_6(k)]^T$ denote the uncertain vectors. The matrices $\mathbf{A}_1, \overline{\mathbf{A}}_1, \mathbf{A}_2, \mathbf{A}_3, \overline{\mathbf{A}}_3, \mathbf{B}_1, \mathbf{B}_2$ and $\mathbf{B}_3$ are defined as follows:

$$\mathbf{A}_1 = \begin{bmatrix} a_{11} & a_{12} \\ a_{21} & a_{22} \end{bmatrix}, \mathbf{A}_2 = \begin{bmatrix} a_{33} & 0 \\ 0 & a_{44} \end{bmatrix}, \mathbf{A}_3 = \begin{bmatrix} a_{51} & a_{52} \\ a_{61} & a_{62} \end{bmatrix}, \overline{\mathbf{A}}_1 = \begin{bmatrix} a_{15} & a_{16} \\ a_{25} & a_{26} \end{bmatrix}$$

$$\overline{\mathbf{A}}_3 = \begin{bmatrix} a_{55} & a_{56} \\ a_{65} & a_{66} \end{bmatrix}, \mathbf{B}_1 = \begin{bmatrix} b_1 & 0 \\ 0 & b_1 \end{bmatrix}, \mathbf{B}_2 = \begin{bmatrix} b_2 & 0 \\ 0 & b_2 \end{bmatrix}, \mathbf{B}_3 = \begin{bmatrix} b_3 & 0 \\ 0 & b_3 \end{bmatrix}$$

3.2.1. Control of Stator Current in the $\alpha - \beta$ Sub-Space

To achieve our control objective, let $\mathbf{x}_1^*(k) = i_{s\phi}^*(k) = \left[i_{s\alpha}^*(k), i_{s\beta}^*(k) \right]^T$ be the vector of desired references with $\phi \in \{\alpha, \beta\}$ and $\mathbf{e}_\phi(k) = \mathbf{x}_1(k) - \mathbf{x}_1^*(k) = i_{s\phi}(k) - i_{s\phi}^*(k)$ be the vector of tracking error. As the relative degree of the stator current in $\alpha - \beta$ sub-space is equal to one, then, the sliding surface [24] is selected to be the error variable as follows:

$$\sigma(k) = \mathbf{e}_\phi(k) \tag{23}$$

In the DSMC design, the following conditions must be satisfied to achieve an ideal sliding motion:

$$\sigma(k) = 0, \quad \sigma(k+1) = 0 \tag{24}$$

where $\sigma(k+1)$ is computed as:

$$\begin{aligned} \sigma(k+1) &= \mathbf{e}_\phi(k+1) = \mathbf{x}_1(k+1) - \mathbf{x}_1^*(k+1) \\ &= \mathbf{A}_1 \mathbf{x}_1(k) + \overline{\mathbf{A}}_1 \mathbf{x}_3(k) + \mathbf{B}_1 \mathbf{u}_1(k) + \eta_1(k) - \mathbf{x}_1^*(k+1) \end{aligned} \tag{25}$$

The control obtained by setting $\sigma(k+1) = 0$ does not ensure robustness and finite-time convergence. For these reasons, the following reaching law is selected:

$$\sigma(k+1) = \Lambda \sigma(k) - T_s \rho \, \text{sign}(\sigma(k)) \tag{26}$$

where $\Lambda = \text{diag}(\lambda_1, \lambda_2)$ with $0 < \lambda_i < 1$ for $i = 1, 2$, $\rho \in R^{2 \times 2}$ is a diagonal positive matrix and $\text{sign}(\sigma(k)) = [\text{sign}(\sigma_1(k)), \text{sign}(\sigma_2(k))]^T$ with:

$$\text{sign}(\sigma_i(k)) = \begin{cases} 1, & \text{if } \sigma_i(k) > 0 \\ 0, & \text{if } \sigma_i(k) = 0 \\ -1, & \text{if } \sigma_i(k) < 0 \end{cases} \tag{27}$$

Then, using (25) and (26), the DSMC law for the stator current in the $\alpha - \beta$ sub-space is obtained as:

$$\mathbf{u}_1(k) = -\mathbf{B}_1^{-1} \left[\mathbf{A}_1 \mathbf{x}_1(k) + \overline{\mathbf{A}}_1 \mathbf{x}_3(k) + \eta_1(k) - \mathbf{x}_1^*(k+1) - \Lambda \sigma(k) + T_s \rho \, \text{sign}(\sigma(k)) \right] \tag{28}$$

The control performance might not be satisfactory since the above equation is in terms of the rotor currents $\mathbf{x}_3(k)$ that are not measurable and the uncertain vector $\eta_1(k)$. Assuming that $\mathbf{x}_3(k)$ and $\eta_1(k)$ do not fluctuate widely between two consecutive sampling times, the TDE method [31,38] can be used to obtain an approximation as:

$$\begin{aligned} \overline{\mathbf{A}}_1 \hat{\mathbf{x}}_3(k) + \hat{\eta}_1(k) &\cong \overline{\mathbf{A}}_1 \mathbf{x}_3(k-1) + \eta_1(k-1) \\ &= \mathbf{x}_1(k) - \mathbf{A}_1 \mathbf{x}_1(k-1) - \mathbf{B}_1 \mathbf{u}_1(k-1) \end{aligned} \tag{29}$$

Definition 1. *For a discrete-time system, a quasi-sliding mode is said to be a trajectory in the vicinity of the sliding surface, such that $|\sigma(k)| < \varepsilon$ and where $\varepsilon > 0$ is the quasi-sliding mode bandwidth. In order to ensure a* convergent quasi-sliding mode, the conditions given in [31,39] that are necessary and sufficient must be verified for each sliding surface, i.e.:

$$\begin{cases} \sigma_i(k) > \varepsilon & \Rightarrow -\varepsilon \leq \sigma_i(k+1) < \sigma_i(k) \\ \sigma_i(k) < -\varepsilon & \Rightarrow \sigma_i(k) < \sigma_i(k+1) \leq \varepsilon \\ |\sigma_i(k)| \leq \varepsilon & \Rightarrow |\sigma_i(k+1)| \leq \varepsilon \end{cases} \tag{30}$$

Theorem 1. *If the following condition is satisfied for i= 1, 2:*

$$\rho_i > \frac{1}{T_s}\delta_i, \tag{31}$$

then, the DSMC with TDE method for the stator currents in the $\alpha - \beta$ sub-space (15) *given by:*

$$\mathbf{u}_1(k) = -\mathbf{B}_1^{-1}\left[\mathbf{A}_1\,\mathbf{x}_1(k) + \overline{\mathbf{A}}_1\,\hat{\mathbf{x}}_3(k) + \hat{\eta}_1(k) - \mathbf{x}_1^*(k+1) - \Lambda\,\sigma(k) - T_s\,\rho\,\mathrm{sign}(\sigma(k))\right] \tag{32}$$

ensures a quasi sliding mode. Moreover, each system trajectory will reach its corresponding sliding surface (23) *within at most $k_i' + 1$ steps, where for i= 1, 2:*

$$k_i' = \frac{|\sigma_i(0)|}{T_s\,\rho_i - \delta_i} \tag{33}$$

Proof of Theorem 1. Substituting the obtained discrete time controller (32) into Equation (25) leads to:

$$\sigma(k+1) = \Lambda\,\sigma(k) + \mathbf{E}(k) - T_s\,\rho\,\mathrm{sign}(\sigma(k)) \tag{34}$$

where $\mathbf{E}(k) = \overline{\mathbf{A}}_1\,(x_3(k) - \hat{x}_3(k)) + (\eta_1(k) - \hat{\eta}_1(k))$ is the bounded TDE error such as for $i = 1,2$:

$$|E_i(k)| < \delta_i \tag{35}$$

Now, choose $\varepsilon = T_s\,\rho_i + \delta_i$. Hence, Equation (30) can be rewritten as:

$$\begin{aligned} \sigma_i(k) > T_s\,\rho_i + \delta_i &\Rightarrow -T_s\,\rho_i - \delta_i \leq \sigma_i(k+1) < \sigma_i(k) \\ \sigma_i(k) < -T_s\,\rho_i - \delta_i &\Rightarrow \sigma_i(k) < \sigma_i(k+1) \leq T_s\,\rho_i + \delta_i \\ |\sigma_i(k)| \leq T_s\,\rho_i + \delta_i &\Rightarrow |\sigma_i(k+1)| \leq T_s\,\rho_i + \delta_i. \end{aligned} \tag{36}$$

1. Consider the first case where $\sigma_i(k) > T_s\,\rho_i + \delta_i$, then $\sigma_i(k) > 0$, $\mathrm{sign}(\sigma_i(k)) = 1$ and:

$$\begin{aligned} \sigma_i(k+1) &= \lambda_i\,\sigma_i(k) + E_i(k) - T_s\,\rho_i \\ \sigma_i(k+1) - \sigma_i(k) &= E_i(k) + (\lambda_i - 1)\,\sigma_i(k) - T_s\,\rho_i. \end{aligned} \tag{37}$$

If the condition in (31) is satisfied, then $\sigma_i(k+1) - \sigma_i(k) < 0 \Rightarrow \sigma_i(k+1) < \sigma_i(k)$.

Moreover, $-T_s\,\rho_i - \delta_i \leq \sigma_i(k+1)$ can be written as:

$$\lambda_i\,\sigma_i(k) + E_i(k) - T_s\,\rho_i \geq -T_s\,\rho_i - \delta_i. \tag{38}$$

Hence:

$$\sigma_i(k) \geq \frac{1}{\lambda_i}\,(E_i(k) - \delta_i), \tag{39}$$

since $\sigma_i(k) > 0$ and $(E_i(k) - \delta_i) < 0$, then the above inequality is always true.

2. Consider the second case where $\sigma_i(k) < -T_s \rho_i - \delta_i$. This implies $\sigma_i(k) < 0$ and $\text{sign}(\sigma_i(k)) = -1$. Then, let us rewrite $\sigma_i(k) < \sigma_i(k+1)$ as follows:

$$\sigma_i(k) < \lambda_i \sigma_i(k) + E_i(k) + T_s \rho_i$$
$$(1 - \lambda_i) \sigma_i(k) < E_i(k) + T_s \rho_i \tag{40}$$

which is always true since $\rho_i > \dfrac{1}{T_s}\delta_i$.

Moreover, $\sigma_i(k+1) < T_s \rho_i + \delta_i$ can be rewritten as:

$$\lambda_i \sigma_i(k) + E_i(k) + T_s \rho_i < T_s \rho_i + \delta_i. \tag{41}$$

Since $\sigma_i(k) < 0$ and $\delta_i > E_i(k)$, then, it is obvious that the inequality in (15) is always true.

3. Consider the third case where $|\sigma_i(k)| \leq \varepsilon$, then:

   a. if $\sigma_i(k) > 0$, then $|\sigma_i(k)| \leq \varepsilon$ becomes:

$$0 < \sigma_i(k) < \varepsilon. \tag{42}$$

Multiplying (42) by λ_i and adding $E_i(k) - T_s \rho_i$ to all the part leads to:

$$E_i(k) - T_s \rho_i < \sigma_i(k+1) < E_i(k) - T_s \rho_i + \lambda_i \varepsilon$$
$$-\varepsilon < \sigma_i(k+1) < \varepsilon \tag{43}$$
$$|\sigma_i(k+1)| \leq \varepsilon$$

   b. if $\sigma_i(k) < 0$, then $|\sigma_i(k)| \leq \varepsilon$ becomes:

$$-\varepsilon < \sigma_i(k) < 0. \tag{44}$$

Once again, multiplying (44) by λ_i and adding $E_i(k) + T_s \rho_i$ to all the parts gives:

$$E_i(k) + T_s \rho_i - \lambda_i \varepsilon < \sigma_i(k+1) < E_i(k) + T_s \rho_i$$
$$-\varepsilon < \sigma_i(k+1) < \varepsilon \tag{45}$$
$$|\sigma_i(k+1)| \leq \varepsilon$$

Hence:

$$|\sigma_i(k+1)| < \varepsilon = T_s \rho_i + \delta_i. \tag{46}$$

Since the conditions in (36) are met, the existence of a convergent quasi sliding mode has been established. Consequently, the proposed DSMC with TDE method in (32) is stable.

Now, let us demonstrate by contradiction according to (34) that Equation (33) is true. For this part, let us assume that $\sigma_i(0) \neq 0$ and $\text{sign}(\sigma_i(0)) = \text{sign}(\sigma_i(1)) = \cdots = \text{sign}(\sigma_i(k'+1))$.

1. Consider the first case where $\sigma_i(0) > 0$ and $\sigma_i(m) > 0$ for all $m \leq (k'_i + 1)$. Then:

$$
\begin{aligned}
\sigma_i(1) &= \lambda_i \sigma_i(0) + E_i(0) - T_s \rho_i \\
&\leq \sigma_i(0) + E_i(0) - T_s \rho_i \\
\sigma_i(2) &\leq \sigma_i(1) + E_i(1) - T_s \rho_i \\
&\leq \sigma_i(0) + E_i(0) + E_i(1) - 2 T_s \rho_i \\
&\vdots \\
\sigma_i(m) &\leq \sigma_i(m-1) + E_i(m-1) - T_s \rho_i \\
&\leq \sigma_i(0) + \sum_{j=0}^{m-1} E_i(j) - m T_s \rho_i \\
&\leq |\sigma_i(0)| + m [\delta_i - T_s \rho_i].
\end{aligned}
\tag{47}
$$

Hence, it is obvious that k'_i ensures that:

$$|\sigma_i(0)| + k'_i [\delta_i - T_s \rho_i] = 0. \tag{48}$$

It follows that:

$$
\begin{aligned}
\sigma_i(k'_i + 1) &\leq |\sigma_i(0)| + (k'_i + 1) [\delta_i - T_s \rho_i] \\
&< |\sigma_i(0)| + k'_i [\delta_i - T_s \rho_i] = 0
\end{aligned}
\tag{49}
$$

which is contradictory to the fact that $\sigma_i(m) > 0$, $\forall m \leq (k'_i + 1)$.

2. Consider the second case where $\sigma_i(0) < 0$ and $\sigma_i(m) < 0$ for all $m \leq (k'_i + 1)$. Then:

$$
\begin{aligned}
\sigma_i(1) &= \lambda_i \sigma_i(0) + E_i(0) + T_s \rho_i \\
&\geq \sigma_i(0) + E_i(0) + T_s \rho_i \\
\sigma_i(2) &\geq \sigma_i(1) + E_i(1) + T_s \rho_i \\
&\geq \sigma_i(0) + E_i(0) + E_i(1) + 2 T_s \rho_i \\
&\vdots \\
\sigma_i(m) &\geq \sigma_i(m-1) + E_i(m-1) + T_s \rho_i \\
&\geq \sigma_i(0) + \sum_{j=0}^{m-1} E_i(j) + m T_s \rho_i \\
&\geq -|\sigma_i(0)| + m [T_s \rho_i - \delta_i]
\end{aligned}
\tag{50}
$$

Once again, it is obvious that k'_i verifies:

$$-|\sigma_i(0)| + k'_i [T_s \rho_i - \delta_i] = 0. \tag{51}$$

It follows that:

$$
\begin{aligned}
\sigma_i(k'_i + 1) &\geq -|\sigma_i(0)| + (k'_i + 1) [T_s \rho_i - \delta_i] \\
&> -|\sigma_i(0)| + k'_i [T_s \rho_i - \delta_i] = 0
\end{aligned}
\tag{52}
$$

which is contradictory to the fact that $\sigma_i(m) < 0$, $\forall m \leq (k'_i + 1)$.

This concludes the proof of Theorem 1. □

3.2.2. Control of Stator Current in the $x - y$ Sub-Space

In this section, the same methodology used previously for the stator current $x_1(k)$ will be adopted to control the stator current in the $x - y$ sub-space. In this case, the sliding surface is selected as follows:

$$\sigma''(k) = \mathbf{e}_{s_{xy}}(k) = \mathbf{x}_2(k) - \mathbf{x}_2^*(k) \tag{53}$$

where $\mathbf{x}_2^*(k) = [i_{sx}^*(k), i_{sy}^*(k)]^T$ is the desired $x - y$ current and $\mathbf{e}_{s_{xy}}(k)$ denotes the tracking error variable. Hence, $\sigma''(k+1)$ is computed as follows:

$$\begin{aligned}\sigma''(k+1) &= \mathbf{e}_{s_{xy}}(k+1) = \mathbf{x}_2(k+1) - \mathbf{x}_2^*(k+1) \\ &= \mathbf{A}_2\,\mathbf{x}_2(k) + \mathbf{B}_2\,\mathbf{u}_2(k) + \eta_2(k) - \mathbf{x}_2^*(k+1).\end{aligned} \tag{54}$$

The discrete-time controller is obtained by solving:

$$\sigma''(k+1) = \Gamma\,\sigma''(k) - T_s\,\varrho\,\mathrm{sign}(\sigma''(k)) \tag{55}$$

where $\Gamma = \mathrm{diag}(\Gamma_1, \Gamma_2)$ with $0 < \Gamma_i < 1$ for $i = 1, 2$, $\varrho \in R^{2 \times 2}$ is a diagonal positive matrix and $\mathrm{sign}(\sigma''(k)) = [\mathrm{sign}(\sigma_1''(k)), \mathrm{sign}(\sigma_2''(k))]^T$, and by substituting the uncertain vector $\eta_2(k)$ by its estimate using TDE method:

$$\begin{aligned}\hat{\eta}_2(k) &\cong \eta_2(k-1) \\ &= \mathbf{x}_2(k) - \mathbf{A}_2\,\mathbf{x}_2(k-1) - \mathbf{B}_2\,\mathbf{u}_2(k-1).\end{aligned} \tag{56}$$

Theorem 2. *If the controller gains are chosen for i = 1, 2 as follows:*

$$\varrho_i > \frac{1}{T_s}\delta_i'' \tag{57}$$

with $\delta_i'' > 0$ the upper-bound of the TDE error $E''(k) = \eta_2(k) - \hat{\eta}_2(k)$, then, the following DSMC with TDE method for the stator current in the $x - y$ sub-space (16) ensures a quasi sliding motion:

$$\mathbf{u}_2(k) = -\mathbf{B}_2^{-1}\left[\mathbf{A}_2\,\mathbf{x}_2(k) + \hat{\eta}_2(k) - \mathbf{x}_2^*(k+1) - \Gamma\,\sigma''(k) + T_s\,\varrho\,\mathrm{sign}(\sigma''(k))\right]. \tag{58}$$

Proof of Theorem 2. The stability analysis is similar to the one described for the stator currents in the $\alpha - \beta$ sub-space. □

4. Experimental Results

The proposed DSMC technique was tested in order to validate its performance with experimental results obtained in the test bench, and this consisted of a six-phase IM powered by two conventional three-phase VSI, being equivalent to a six-leg VSI, using a constant DC-bus voltage from a DC power supply system. The six-leg VSI was controlled by a dSPACE MABXII DS1401 real-time rapid prototyping platform, with Simulink version 8.2. The results obtained were captured and processed using MATLAB R2013b script. The parameters of the asymmetrical six-phase IM were obtained using conventional methods of the AC time domain and stand-still with VSI supply tests [40,41]. The results are listed in Table 1. The experimental tests were performed with current sensors LA 55-P s, which had a frequency bandwidth from DC up to 200 kHz. The current measurements were then converted to digital form using a 16-bit A/D converter. The six-phase IM position was obtained with a 1024-ppr incremental encoder, and the rotor speed was estimated from it. Finally, a 5 HP eddy current brake was used to introduce a variable mechanical load on the IM. A block diagram of the experimental bench is shown in Figure 3, including some photos of the equipment.

Table 1. Parameters of the six-phase IM.

Parameter	Value	Parameter	Value	Parameter	Value
R_r (Ω)	6.9	L_r (mH)	626.8	P_w (kW)	2
R_s (Ω)	6.7	ω_{m-nom} (rpm)	3000	J_i (kg·m^2)	0.07
L_{ls} (mH)	5.3	L_s (mH)	654.4	B_i (kg·m^2/s)	0.0004
L_m (mH)	614	P	1	V_{dc} (V)	400

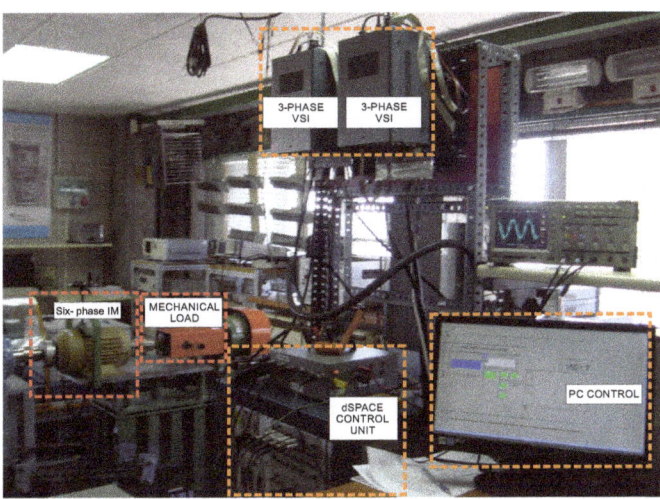

Figure 3. Block diagram of the test bench including the six-phase IM, the six-leg VSI, the dSPACE and the mechanical load.

The performance of the proposed DSMC was analysed in transient and steady-state conditions. The experimental results analysed the controller performance in terms of Mean Squared Error (MSE) between the reference and measured stator currents in the $\alpha - \beta$, $x - y$ and $d - q$ planes. The Root Mean Square (RMS) of the currents in the $d - q$ plane was used to calculate their corresponding Form Factor (FF) and Total Harmonic Distortion (THD) obtained in the $\alpha - \beta$ plane, as well as MSE for rotor speed. The MSE is defined as:

$$\text{MSE}(i_{s\Phi}) = \sqrt{\frac{1}{N} \sum_{k=1}^{N} (i_{s\Phi}(k) - i_{s\Phi}^*(k))^2} \tag{59}$$

where N is the number of analysed samples, $i_{s\Phi}^*$ the stator current reference, $i_{s\Phi}$ the measured stator currents and $\Phi \in \{\alpha, \beta, x, y, d, q\}$. On the other hand, the THD is calculated as:

$$\text{THD}(i_s) = \sqrt{\frac{1}{i_{s1}^2} \sum_{j=2}^{N} (i_{sj})^2} \tag{60}$$

where i_{s1} is the fundamental stator currents and i_{sj} is the harmonic stator currents. At last, the FF is computed as:

$$\text{FF}(i_{dqs}) = \frac{i_{dqs-RMS}}{i_{dqs-mean}}. \tag{61}$$

A fixed d current ($i_{ds}^* = 1$ A) was used. To perform a mechanical load for the six-phase IM, the eddy current brake was fixed at 1.65 A. Moreover, the chosen gains of the DSMC with TDE for stator current tracking are:

$$\lambda = \text{diag}(0.5, 0.5), \qquad \rho_1 = \rho_2 = 100,$$

$$\Gamma = \text{diag}(0.9, 0.9), \qquad \varrho_1 = \varrho_2 = 100.$$

The stator current reference in the $x - y$ sub-space was set to zero ($i_{xs}^* = i_{ys}^* = 0$ A) in order to reduce the copper losses. The sampling frequencies used in the tests were 8 kHz and 16 kHz. Three operation points were set for the rotor speed: 500 rpm, 1000 rpm and 1500 rpm for steady-state analysis. For a transient response, a step change in rotor speed was considered from 500 to -500 rpm (i.e., a reversal condition).

The proposed technique DSMC was tested under different operating points in steady state and under transient conditions. Table 2 shows the experimental results obtained for different rotor mechanical speeds and sampling frequencies, regarding the MSE of stator currents in the $\alpha - \beta$, $x - y$ and $d - q$ planes. The results showed good performance of DSMC applied to the six-phase IM in terms of current tracking, in their corresponding planes, especially in $\alpha - \beta$ current tracking. Table 3 shows the results of THD in $\alpha - \beta$ stator currents, RMS ripple and FF in $d - q$ currents and the MSE of the measured and referenced rotor speed. The results presented a reduction on the THD stator currents with the higher sampling frequency and higher rotor speed. In terms of RMS ripple and FF, there was a significant reduction with higher sampling frequency in all the rotor speed tests. However, for rotor speed MSE, better performance was obtained at lower rotor speed and sampling frequency, but this was not significant.

Figure 4 presents the polar trajectories of stator currents in the $x - y$ and $\alpha - \beta$ sub-spaces at different rotor speeds. The tests were developed with the same mechanical load; thus, the amplitude of $\alpha - \beta$ currents was proportional to the rotor speed. The figures show that $x - y$ currents were reduced to almost the same ratio in every case and $\alpha - \beta$ current tracking was good. On the other hand, Figures 5 and 6 report a dynamic test, which consisted of the transient performance of DSMC for a step response in the q axis current reference (i_{qs}^*). The dynamic response was generated through a reversal condition of the rotor mechanical speed (ω_m) from 500 to -500 rpm. Figures 5a and 6a show an overshoot of 42% and 70%, respectively, and a settling time of 1.3 ms and 1.4 ms. respectively, presenting in both cases very fast responses.

Table 2. Performance analysis of stator currents $\alpha - \beta$, $x - y$, $d - q$ and MSE (A) for three different rotor speeds (rpm).

	Sampling	Frequency	8 kHz			
ω_m^*	MSE_α	MSE_β	MSE_x	MSE_y	MSE_d	MSE_q
500	0.2502	0.2602	0.1875	0.1729	0.2494	0.2609
1000	0.2937	0.3021	0.2326	0.2280	0.3039	0.2919
1500	0.3000	0.3050	0.2491	0.2456	0.3327	0.2689
	Sampling	Frequency	16 kHz			
ω_m^*	MSE_α	MSE_β	MSE_x	MSE_y	MSE_d	MSE_q
500	0.1867	0.1883	0.1931	0.1851	0.1830	0.1919
1000	0.1797	0.1779	0.2078	0.1975	0.1795	0.1780
1500	0.1731	0.1786	0.2342	0.2291	0.1767	0.1750

Table 3. Performance analysis of stator current $\alpha - \beta$, THD (%), $d - q$, RMS ripple (A), FF, rotor speed (ω_m) and MSE (rpm) at different rotor speeds (rpm).

	Sampling	Frequency	8 kHz				
ω_m^*	THD_α	THD_β	RMS ripple$_q$	RMS ripple$_d$	FF_q	FF_d	MSE_{ω_m}
500	29.6198	30.7074	0.2598	0.2492	1.0811	1.0300	1.3432
1000	17.8543	18.0026	0.2890	0.3005	1.0203	1.0405	2.2250
1500	17.8761	18.0059	0.2593	0.3194	1.0084	1.1389	2.4146
	Sampling	Frequency	16 kHz				
ω_m^*	THD_α	THD_β	RMS ripple$_q$	RMS ripple$_d$	FF_q	FF_d	MSE_{ω_m}
500	21.6914	22.6592	0.1895	0.1829	1.0466	1.0164	1.6508
1000	15.3291	14.8507	0.1751	0.1783	1.0087	1.0151	2.8814
1500	11.1020	11.2140	0.1707	0.1712	1.0040	1.0134	3.1855

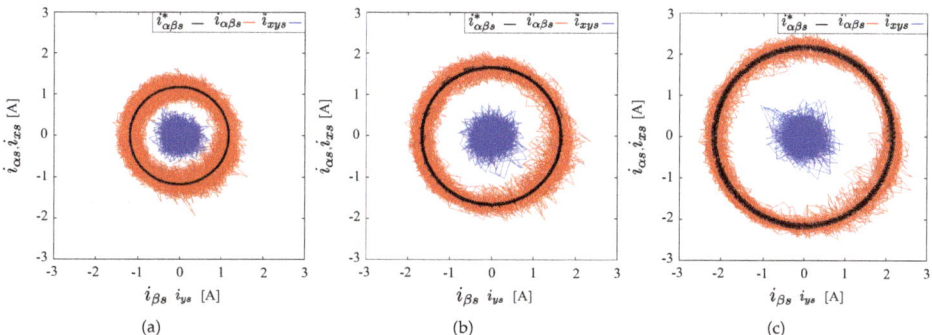

Figure 4. Stator currents in the $\alpha - \beta$ and $x - y$ planes for a rotor speed ω_m of: (**a**) 500 rpm; (**b**) 1 000 rpm; (**c**) 1 500 rpm.

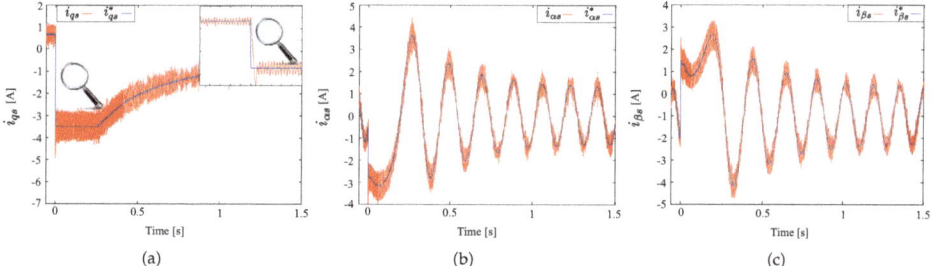

Figure 5. Transient response of stator currents from a step response of 500 rpm to -500 rpm from ω_m at a frequency sample of 8 kHz: (**a**) i_{qs}; (**b**) $i_{\alpha s}$; (**c**) $i_{\beta s}$.

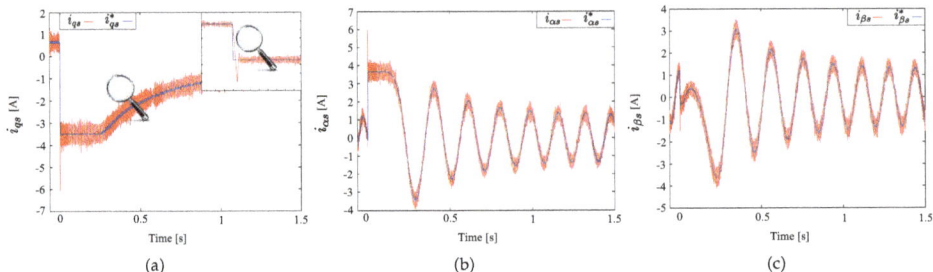

Figure 6. Transient response of stator currents from a step response of 500 rpm to −500 rpm from ω_m at a frequency sample of 16 kHz: (**a**) i_{qs}; (**b**) $i_{\alpha s}$; (**c**) $i_{\beta s}$.

5. Conclusions

In this work, a speed control based on the IRFOC strategy with an inner robust DSMC with the TDE method for stator currents in the $\alpha - \beta$ and $x - y$ sub-spaces has been proposed. On the one hand, the TDE method allows in a simple way highly accurate estimation of the uncertainties, perturbations and unmeasurable rotor current. On the other hand, discrete-time sliding mode cancels the effect of the TDE error, ensures robustness and delivers high precision and fast convergence. The efficiency of the proposed discrete control scheme has been confirmed by a real-time implementation on a six-phase induction motor drive. The proposed approach provides very good performances in dynamic processes, as well as in steady state. Moreover, the average switching frequency of the designed DSMC is low. Further research will be initiated to benefit from the advantages offered by multiphase machines. To that end, an extension of the proposed controller will be developed in the case of an open circuit fault in one or more phases occurring, since this fault is common for induction machines. The work will focus on the ability of ensuring good performances without good knowledge of the new mathematical model of the machine under fault condition.

Author Contributions: Conceptualization, Y.K., J.R. and M.S.; methodology, Y.K., M.A. and J.R.; software, Y.K., M.A. and J.R.; validation, M.A. and J.D.-G.; formal analysis, Y.K., J.R., M.S. and K.B.; investigation, Y.K. and J.R.; resources, J.D.-G.; data curation, J.D.-G. and M.A.; writing, original draft preparation, Y.K., M.A. and J.R.; writing, review and editing, Y.K., M.A., J.R., M.S., J.D.-G. and R.G.; visualization, J.D.-G., M.A. and J.R.; project administration, J.R., M.S. and J.D.-G.; funding acquisition, R.G. and J.R.

Funding: This research and APC were funded by the Consejo Nacional de Ciencia y Tecnología (CONACYT)-Paraguay, Grant Numbers 14-INV-101 and POSG16-05.

Acknowledgments: The authors would like to thank Graham Goodwin (University of Newcastle-Australia) for his valuable comments on this research work.

Conflicts of Interest: The authors declare no conflict of interest.

Abbreviations

The following abbreviations are used in this manuscript:

DSMC	Discrete-Time Siding Mode Control
FF	Form Factor
IM	Induction Machine
IRFOC	Indirect Rotor Field-Oriented Control
MSE	Mean Squared Error
RMS	Root Mean Square
PI	Proportional-Integral
SMC	Sliding Mode Control
TDE	Time Delay Estimation
THD	Total Harmonic Distortion

VSD   Vector Space Decomposition
VSI   Voltage Source Inverter

References

1. Barrero, F.; Duran, M.J. Recent Advances in the Design, Modeling, and Control of Multiphase Machines: Part I. *IEEE Trans. Ind. Electron.* **2016**, *63*, 449–458. [CrossRef]
2. Duran, M.J.; Barrero, F. Recent Advances in the Design, Modeling, and Control of Multiphase Machines: Part II. *IEEE Trans. Ind. Electron.* **2016**, *63*, 459–468. [CrossRef]
3. Levi, E. Advances in Converter Control and Innovative Exploitation of Additional Degrees of Freedom for Multiphase Machines. *IEEE Trans. Ind. Electron.* **2016**, *63*, 433–448. [CrossRef]
4. Zoric, I.; Jones, M.; Levi, E. Arbitrary Power Sharing Among Three-Phase Winding Sets of Multiphase Machines. *IEEE Trans. Ind. Electron.* **2018**, *65*, 1128–1139. [CrossRef]
5. Yepes, A.G.; Malvar, J.; Vidal, A.; López, O.; Doval-Gandoy, J. Current harmonics compensation based on multiresonant control in synchronous frames for symmetrical n-phase machines. *IEEE Trans. Ind. Electron.* **2015**, *62*, 2708–2720. [CrossRef]
6. Lim, C.; Levi, E.; Jones, M.; Rahim, N.; Hew, W.P. FCS-MPC based current control of a five-phase induction motor and its comparison with PI-PWM control. *IEEE Trans. Ind. Electron.* **2014**, *61*, 149–163. [CrossRef]
7. Taheri, A.; Rahmati, A.; Kaboli, S. Efficiency improvement in DTC of six-phase induction machine by adaptive gradient descent of flux. *IEEE Trans. Power Electron.* **2012**, *27*, 1552–1562. [CrossRef]
8. Riveros, J.A.; Barrero, F.; Levi, E.; Duran, M.J.; Toral, S.; Jones, M. Variable-speed five-phase induction motor drive based on predictive torque control. *IEEE Trans. Ind. Electron.* **2013**, *60*, 2957–2968. [CrossRef]
9. Gregor, R.; Rodas, J. Speed sensorless control of dual three-phase induction machine based on a Luenberger observer for rotor current estimation. In Proceedings of the 38th Annual Conference on IEEE Industrial Electronics Society (IECON), Montreal, QC, Canada, 25–28 Octorber 2012; pp. 3653–3658. [CrossRef]
10. Ayala, M.; Gonzalez, O.; Rodas, J.; Gregor, R.; Doval-Gandoy, J. A speed-sensorless predictive current control of multiphase induction machines using a Kalman filter for rotor current estimator. In Proceedings of the 2016 International Conference on Electrical Systems for Aircraft, Railway, Ship Propulsion and Road Vehicles & International Transportation Electrification Conference (ESARS-ITEC), Toulouse, France, 2–4 November 2016; pp. 1–6. [CrossRef]
11. Rodas, J.; Barrero, F.; Arahal, M.R.; Martín, C.; Gregor, R. Online estimation of rotor variables in predictive current controllers: a case study using five-phase induction machines. *IEEE Trans. Ind. Electron.* **2016**, *63*, 5348–5356. [CrossRef]
12. Rodas, J.; Martín, C.; Arahal, M.R.; Barrero, F.; Gregor, R. Influence of Covariance-Based ALS Methods in the Performance of Predictive Controllers with Rotor Current Estimation. *IEEE Trans. Ind. Electron.* **2017**, *64*, 2602–2607. [CrossRef]
13. Guzman, H.; Duran, M.J.; Barrero, F.; Zarri, L.; Bogado, B.; Prieto, I.G.; Arahal, M.R. Comparative study of predictive and resonant controllers in fault-tolerant five-phase induction motor drives. *IEEE Trans. Ind. Electron.* **2016**, *63*, 606–617. [CrossRef]
14. Bermudez, M.; Gonzalez-Prieto, I.; Barrero, F.; Guzman, H.; Duran, M.J.; Kestelyn, X. Open-Phase Fault-Tolerant Direct Torque Control Technique for Five-Phase Induction Motor Drives. *IEEE Trans. Ind. Electron.* **2017**, *64*, 902–911. [CrossRef]
15. Baneira, F.; Doval-Gandoy, J.; Yepes, A.G.; López, O.; Pérez-Estévez, D. Control Strategy for Multiphase Drives With Minimum Losses in the Full Torque Operation Range Under Single Open-Phase Fault. *IEEE Trans. Power Electron.* **2017**, *32*, 6275–6285. [CrossRef]
16. Rodas, J.; Guzman, H.; Gregor, R.; Barrero, B. Model predictive current controller using Kalman filter for fault-tolerant five-phase wind energy conversion systems. In Proceedings of the 7th International Symposium on Power Electronics for Distributed Generation Systems (PEDG), Vancouver, BC, Canada, 27–30 June 2016; pp. 1–6. [CrossRef]
17. Echeikh, H.; Trabelsi, R.; Mimouni, M.F.; Iqbal, A.; Alammari, R. High performance backstepping control of a fivephase induction motor drive. In Proceedings of the 2014 IEEE 23rd International Symposium on Industrial Electronics (ISIE), Istanbul, Turkey, 1–4 June 2014; pp. 812–817. [CrossRef]

18. Echeikh, H.; Trabelsi, R.; Iqbal, A.; Bianchi, N.; Mimouni, M.F. Comparative study between the rotor flux oriented control and non-linear backstepping control of a five-phase induction motor drive—An experimental validation. *IET Power Electron.* **2016**, *9*, 2510–2521. [CrossRef]
19. Kali, Y.; Rodas, J.; Saad, M.; Gregor, R.; Bejelloun, K.; Doval-Gandoy, J. Current Control based on Super-Twisting Algorithm with Time Delay Estimation for a Five-Phase Induction Motor Drive. In Proceedings of the 2017 International Electric Machines and Drives Conference (IEMDC), Miami, FL, USA, 21–24 May 2017; pp. 1–8. [CrossRef]
20. Kali, Y.; Rodas, J.; Saad, M.; Gregor, R.; Bejelloun, K.; Doval-Gandoy, J.; Goodwin, G. Speed Control of a Five-Phase Induction Motor Drive using Modified Super-Twisting Algorithm. In Proceedings of the 2018 International Symposium on Power Electronics, Electrical Drives, Automation and Motion (SPEEDAM), Amalfi, Italy, 20–22 June 2018; pp. 938–943. [CrossRef]
21. Ayala, M.; Gonzalez, O.; Rodas, J.; Gregor, R.; Kali, Y.; Wheeler, P. Comparative Study of Non-linear Controllers Applied to a Six-Phase Induction Machine. In Proceedings of the 2018 International Conference on Electrical Systems for Aircraft, Railway, Ship Propulsion and Road Vehicles & International Transportation Electrification Conference (ESARS-ITEC), Nottingham, UK, 7–9 November 2018; pp. 1–6.
22. Fnaiech, M.A.; Betin, F.; Capolino, G.A.; Fnaiech, F. Fuzzy logic and sliding-mode controls applied to six-phase induction machine with open phases. *IEEE Trans. Ind. Electron.* **2010**, *57*, 354–364. [CrossRef]
23. Utkin, V. *Sliding Mode in Control and Optimization*; Springer-Verlag: Berlin, German, 1992.
24. Utkin, V.; Guldner, J.; Shi, J. *Sliding Mode Control in Electromechanical Systems*; Taylor-Francis: Abingdon, UK, 1999.
25. Fridman, L. An averaging approach to chattering. *IEEE Trans. Autom. Control* **2001**, *46*, 1260–1265. [CrossRef]
26. Boiko, I.; Fridman, L. Analysis of Chattering in Continuous Sliding-mode Controllers. *IEEE Trans. Autom. Control* **2005**, *50*, 1442–1446. [CrossRef]
27. Young, K.D.; Utkin, V.I.; Ozguner, U. A control engineer's guide to sliding mode control. *IEEE Trans. Control Syst. Technol.* **1999**, *7*, 328–342. [CrossRef]
28. Drakunov, S.; Utkin, V. Sliding mode observers. Tutorial. In Proceedings of the 34th IEEE Conference on Decision and Control, New Orleans, LA, USA, 13–15 December 1995; pp. 3376–3378. [CrossRef]
29. Yan, X.G.; Edwards, C. Nonlinear robust fault reconstruction and estimation using a sliding mode observer. *Automatica* **2007**, *43*, 1605–1614. [CrossRef]
30. Levant, A. Higher-order sliding modes, differentiation and output-feedback control. *Int. J. Control* **2003**, *76*, 924–941. [CrossRef]
31. Kali, Y.; Saad, M.; Benjelloun, K.; Fatemi, A. Discrete-time second order sliding mode with time delay control for uncertain robot manipulators. *Robot. Auton. Syst.* **2017**, *94*, 53–60. [CrossRef]
32. Kali, Y.; Saad, M.; Benjelloun, K.; Khairallah, C. Super-twisting algorithm with time delay estimation for uncertain robot manipulators. *Nonlinear Dyn.* **2018**, *93*, 557–569. [CrossRef]
33. Kali, Y.; Saad, M.; Benjelloun, K.; Benbrahim, M. Sliding Mode with Time Delay Control for MIMO Nonlinear Systems with Unknown Dynamics. In Proceedings of the 2015 International Workshop on Recent Advances in Sliding Modes (RASM), Istanbul, Turkey, 9–11 April 2015; pp. 1–6. [CrossRef]
34. Kali, Y.; Rodas, J.; Gregor, R.; Saad, M.; Benjelloun, K. Attitude Tracking of a Tri-Rotor UAV based on Robust Sliding Mode with Time Delay Estimation. In Proceedings of the 2018 International Conference on Unmanned Aircraft Systems (ICUAS), Dallas, TX, USA, 12–15 June 2018; pp. 346–351. [CrossRef]
35. Bandal, V.; Bandyopadhyay, B.; Kulkarni, A.M. Design of power system stabilizer using power rate reaching law based sliding mode control technique. In Proceedings of the 2005 International Power Engineering Conference, Singapore, 29 November–2 December 2005; pp. 923–928. [CrossRef]
36. Gonzalez, O.; Ayala, M.; Rodas, J.; Gregor, R.; Rivas, G.; Doval-Gandoy, J. Variable-Speed Control of a Six-Phase Induction Machine using Predictive-Fixed Switching Frequency Current Control Techniques. In Proceedings of the 9th IEEE International Symposium on Power Electronics for Distributed Generation Systems (PEDG), Charlotte, NC, USA, 25–28 June 2018; pp. 1–6. [CrossRef]
37. Harnefors, L.; Saarakkala, S.E.; Hinkkanen, M. Speed Control of Electrical Drives Using Classical Control Methods. *IEEE Trans. Ind. Appl.* **2013**, *49*, 889–898. [CrossRef]
38. Jung, J.H.; Chang, P.; Kang, S.H. Stability Analysis of Discrete Time Delay Control for Nonlinear Systems. In Proceedings of the 2007 American Control Conference, New York City, NY, USA, July 11–13 2007; pp. 5995–6002. [CrossRef]

39. Qu, S.; Xia, X.; Zhang, J. Dynamics of Discrete-Time Sliding-Mode-Control Uncertain Systems With a Disturbance Compensator. *IEEE Trans. Ind. Electron.* **2014**, *61*, 3502–3510. [CrossRef]
40. Yepes, A.G.; Riveros, J.A.; Doval-Gandoy, J.; Barrero, F.; López, O.; Bogado, B.; Jones, M.; Levi, E. Parameter identification of multiphase induction machines with distributed windings Part 1: Sinusoidal excitation methods. *IEEE Trans. Energy Convers.* **2012**, *27*, 1056–1066. [CrossRef]
41. Riveros, J.A.; Yepes, A.G.; Barrero, F.; Doval-Gandoy, J.; Bogado, B.; Lopez, O.; Jones, M.; Levi, E. Parameter identification of multiphase induction machines with distributed windings Part 2: Time-domain techniques. *IEEE Trans. Energy Convers.* **2012**, *27*, 1067–1077. [CrossRef]

© 2019 by the authors. Licensee MDPI, Basel, Switzerland. This article is an open access article distributed under the terms and conditions of the Creative Commons Attribution (CC BY) license (http://creativecommons.org/licenses/by/4.0/).

Article

Interest and Applicability of Meta-Heuristic Algorithms in the Electrical Parameter Identification of Multiphase Machines [†]

Daniel Gutierrez-Reina [1], Federico Barrero [2,*], Jose Riveros [3], Ignacio Gonzalez-Prieto [4], Sergio L. Toral [2] and Mario J. Duran [5]

[1] Department of Engineering, Loyola University Andalusia, 41014 Seville, Spain; dgutierrez@uloyola.es
[2] Electronic Engineering Department, University of Seville, 41092 Sevilla, Spain; storal@us.es
[3] Faculty of Engineering, University of Talca, Curicó 3340000, Chile; joservs@gmail.com
[4] Thermal and Electrical Engineering Department, University of Huelva, 21007 Huelva, Spain; ignacio.gonzalez@die.uhu.es
[5] Department of Electrical Engineering, University of Malaga, 29071 Malaga, Spain; mjduran@uma.es
* Correspondence: fbarrero@us.es; Tel.: +34-954481304
[†] This paper is an extended version of our paper published in Riveros, J.A.; Reina, D.G.; Barrero, F.; Toral, S.L.; Durán, M.J. Five-Phase Induction Machine Parameter Identification using PSO and Standstill Techniques. In Proceedings of the IECON 2015—41st Annual Conference of the IEEE Industrial Electronics Society, Yokohama, Japan, 9–12 November 2015.

Received: 26 November 2018; Accepted: 16 January 2019; Published: 19 January 2019

Abstract: Multiphase machines are complex multi-variable electro-mechanical systems that are receiving special attention from industry due to their better fault tolerance and power-per-phase splitting characteristics compared with conventional three-phase machines. Their utility and interest are restricted to the definition of high-performance controllers, which strongly depends on the knowledge of the electrical parameters used in the multiphase machine model. This work presents the proof-of-concept of a new method based on particle swarm optimization and standstill time-domain tests. This proposed method is tested to estimate the electrical parameters of a five-phase induction machine. A reduction of the estimation error higher than 2.5% is obtained compared with gradient-based approaches.

Keywords: multiphase drives; off-line identification methods; meta-heuristic algorithms

1. Introduction

Electromechanical systems such as multiphase variable speed drives have attracted the interest of the scientific community in recent times. They have been found as an attractive alternative to three-phase drives in particular industrial applications [1], where the electrical stresses on the machine and power electronic components as well as the harmonic content must be reduced and/or an inherent fault-tolerant capability is required. The interest in recent research works aims to exploit the inherent characteristics of multiphase drives, improving the overall reliability and performance of the system in order to favor their industrial applicability. However, their higher number of phases, in comparison with three-phase drives, results in more complex controllers due to higher number of freedom degrees. Most of the control techniques that have been proposed for multiphase drives are an extension of conventional three-phase control structures, aiming for a high speed/torque performance of the drive in healthy and faulty situations, and giving particular attention to multiphase machines of five and six phases [1,2]. Then, field oriented control (FOC) techniques, direct torque controllers (DTCs) or model-based predictive control (MPC) methods have been successfully used in multiphase drives, where an accurate knowledge of the electrical parameters of the machine is required to yield the highest performance behavior of a system [1,2]. Note however that multiphase drives can be considered like an emerging technology, where most existing units have been built by rewinding conventional three-phase machines and reshaping the distribution of the stator slots [3,4]. The resulting machine is neither the most optimal nor its parameters correspond with those of the original three-phase drive. Therefore, methods and algorithms for the estimation of the rewound machine's parameters are required to get adjustable speed multiphase drives with appropriate control performances.

While the research on the identification of the electrical parameters of conventional three-phase drives is a mature field, this is not the case in the multiphase drives' area [1]. Many off-line and on-line methods have been proposed to obtain the electrical parameters of three-phase machines, where standstill identification techniques can be highlighted for being accurate and easy to apply in commercial variable frequency drives [5,6]. Standstill methods are off-line identification tools based on injecting dc or ac electrical signals using the power converter of the drive, normally a Voltage Source Inverter (VSI), which does not produce a rotating field and keeps the electrical machine stopped. Then, the identification procedure is applied to fit the real response with the simplified machine model, where adaptive filters, recursive least-squares (RLS)-based algorithms, or maximum likelihood methods have been used for this purpose [7]. The extension of these methods for the multiphase case is barely found in the scientific literature. The standstill methods have been successfully applied for the identification of the electrical parameters of a symmetrical 5-phase induction machine with distributed windings in [8,9]. In [8,9], the stator and rotor resistors, the mutual inductance and the stator and rotor leakage inductances of the machine modelling are estimated using the non-torque capability of particular harmonic components that are injected in the estimation process. A RLS procedure was applied to fit the real response with the machine model, also complemented with sinusoidal excitation methods to tune and adjust the estimated parameters. The obtained results offer however bad accuracy and high deviation in some trials (up to 50% for certain cases in the estimation of the magnetizing inductance) because it is based on gradient-following-based algorithms that cannot properly fit the non-linear performance of a real machine. The algorithm proposed in [8,9] shows also a high dependency on the established forgetting factors, requiring an initial value for the estimated parameters close to the optimum result to find the global minimum solution. In this work, the method in [8,9] is extended to find an identification scheme that avoids the aforementioned drawbacks and propagation errors, adding the ability of detecting constructive asymmetries in the machine if desired.

Meta-heuristic algorithms may represent an interesting alternative in this field [10]. These methods can offer a suitable guided search even in non-differentiable or nonlinear spaces, where conventional gradient-based methods are usually unsuccessful [11] because they get stuck in local minima. Among the available meta-heuristic optimization techniques, the particle swarm optimization (PSO) algorithm [12,13] is an interesting tool for solving optimization complex engineering problems [14,15]. It is based on the metaphor of social interaction during the movement into a multidimensional space and it has been widely applied for solving power systems optimization problems [14]. In this paper, the PSO optimization technique is proposed to minimize the mean square error (MSE) in two operation subspaces, namely α–β and x–y, between the responses of the simulated and real systems in standstill configuration for a multi-variable electro-mechanical system like a five-phase induction machine. To the authors' knowledge, this is the first study that applies a bio-inspired algorithm like the PSO for the estimation of electrical parameters in electro-mechanical systems. The main idea is that the simulated model reaches the same responses of the real system, as the estimated parameters get closer to the real ones guided by the PSO algorithm.

Therefore, the main contributions of this paper are:

- The analysis of the utility of PSO algorithms in an application-oriented case like the estimation of the electrical parameters of a five-phase induction machine.
- The comparison of the proposed PSO estimation technique with gradient-following-based algorithms [16]. The proposed technique clearly outperforms the gradient-based technique.

This paper continues as follows: Section 2 overviews the five-phase induction drives, which is the multi-variable electro-mechanical system used as case example and the PSO algorithm. Section 3 analyses the proposed estimation procedure that combines standstill tests and the PSO algorithm. Section 4 provides the estimated electrical parameters achieved by the proposed method and the validation of the obtained parameters in a real test. Finally, conclusions are given in Section 5.

2. Background: Five-Phase Induction Machines and PSO Algorithm

This section is divided into two parts. First, an introduction of the five-phase induction machine used in the paper is presented. Second, the PSO algorithm used and its configuration parameters are described in details.

2.1. Five-Phase Induction Machines

The case under study is a symmetrical five-phase induction machine, where the stator windings are equally displaced ($\vartheta = 2\pi/5$) and sinusoidally distributed along the stator. The multiphase drive is power-supplied using a two-level VSI, as can be observed in Figure 1.

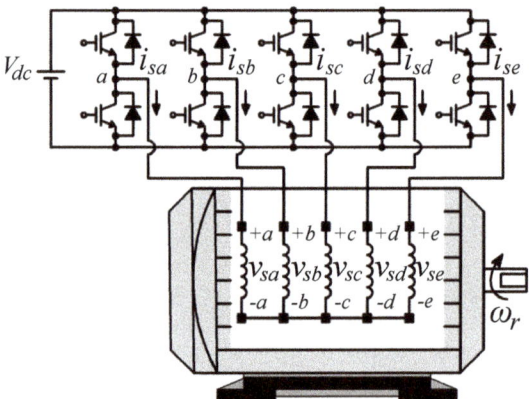

Figure 1. General scheme of the system under study.

The model of the system is more complex than the one obtained for a three-phase case due to the higher number of phases. However, the general theory of electrical machines is also applied to obtain the model of the system and the following assumptions are taken into account to obtain a set of continuous-time phase voltage equilibrium equations: machine windings are identical and equally distributed around the stator, magnetic field saturation and eddy currents are not considered, non-linearity in relation with temperature or frequency changes are not considered, and the machine air gap is assumed to be uniform and of constant density without any variation due to rotor eccentricities or machine slots. These equations can be simplified to avoid the dependence of the rotor position of certain parameter matrices using the Clarke transformation, which is used by the vector space decomposition theory to determine two orthogonal planes completely decoupled from each other (called α-β and x-y), plus an axis that contains the homopolar component (z-component). The obtained equations are detailed in (1)–(7), where the electrical parameters to be estimated are shown (the stator and rotor resistances, R_s and R_r, respectively, the mutual inductance represented by L_m, and the stator and rotor leakage inductances, L_{ls} and L_{lr}, respectively). It is interesting to mention that the fundamental supply component plus harmonics of the order $10n \pm 1$ ($n = 0,1,2,3,...$) are within the α–β subspace, which is the torque-producing plane. The rest of harmonic components are into the non-torque producing planes, including the x–y subspace, where supply harmonics of the order $10n \pm 3$ ($n = 0,1,2,3,...$) are considered, and the z-axis that contains harmonic components of the order $5n$, with $n = 1,2,3,...$ and only exists if the neutral point is not isolated. This is not our case because isolated neutral point is assumed and (7) is no longer required because $i_{sz} = 0$. Therefore, 32 (2^5) switching states and 30 active, and 2 zero voltage vectors can be generated in the α-β and x-y subspaces. Figure 2 identifies all available voltage vectors that can be applied to the multiphase machine, identified by using the decimal number corresponding to the binary code of the switching state S_a, S_b, S_c, S_d, S_e, being S_a and S_e the most and least significant bits, respectively. The modelling of the machine is finally complemented with a differential equation that describes the rotor movement depending on the electrical and load torques. Since this study focuses on the estimation of the electrical parameters of the machine, the movement equation is omitted here for simplicity (more details on the modelling of system can be found in [1–4]):

$$v_{s\alpha} = \left(R_s + L_s\frac{d}{dt}\right)i_{s\alpha} + L_m\frac{di_{r\alpha}}{dt} \tag{1}$$

$$v_{s\beta} = \left(R_s + L_s\frac{d}{dt}\right)i_{s\beta} + L_m\frac{di_{r\beta}}{dt} \tag{2}$$

$$0 = \left(R_r + L_r\frac{d}{dt}\right)i_{r\alpha} + L_m\frac{di_{s\alpha}}{dt} + \omega_r L_r i_{r\beta} + \omega_r L_m i_{s\beta} \tag{3}$$

$$0 = \left(R_r + L_r\frac{d}{dt}\right)i_{r\beta} + L_m\frac{di_{s\beta}}{dt} - \omega_r L_r i_{r\alpha} - \omega_r L_m i_{s\alpha} \tag{4}$$

$$v_{sx} = \left(R_s + L_{ls}\frac{d}{dt}\right)i_{sx} \tag{5}$$

$$v_{sy} = \left(R_s + L_{ls}\frac{d}{dt}\right)i_{sy} \tag{6}$$

$$v_{sz} = \left(R_s + L_{ls}\frac{d}{dt}\right)i_{sz} \tag{7}$$

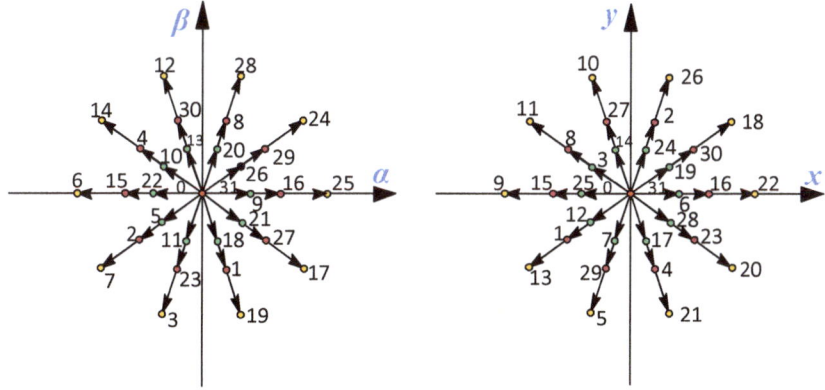

Figure 2. Generated voltage vectors in the α-β and x-y planes. Note that the same switching state produces two different vectors in every plane.

2.2. PSO Algorithm

PSO is a meta-heuristic population-based technique. It is inspired by the social behavior of bird flocking and fish schooling; therefore, it is based on the swarm intelligence concept [17]. PSO refers to artificial intelligence systems where the collective behavior of unsophisticated agents that interact locally with their environment creates coherent global functional patterns [12,15]. In general,

PSO algorithm uses a population of particles that fly throughout the problem hyperspace [18]. All the particles have fitness values that are evaluated by the fitness function to be optimized and have velocities vectors, which determine the movement of the particles in the search space. These velocities are stochastically adjusted throughout the execution of the algorithm according to the historical best position for the particle itself and the neighborhood (other neighbor particles) [12,15]. Therefore, the particles or candidate solutions fly throughout the problem search space attracted by the positions of the best particles found during the execution of the algorithm. PSO-based methods have been used in a wide range of engineering areas to solve complex continuous optimization problems, such as product design and manufacturing [19], automotive industry [20], structural design [21], and computer networks [22], among others [23,24].

Mathematically, the PSO algorithm is formulated as follows. First, a set of P particles (population) is randomly initialized. Note that the position of each particle is a possible solution for the estimation algorithm and it is represented by a d-dimensional vector in the problem space $x_i = (x_{i1}, x_{i2}, \ldots, x)$, being $i = 1, 2, \ldots, P$ and $s \in \mathbb{R}$. Thus, each particle is randomly placed in the d-dimensional space as a candidate solution and its performance is evaluated using a predefined fitness function. The velocity of the ith particle $v_i = v_{i1}, v_{i2}, \ldots, v_{id}$, $v \in \mathbb{R}$, is defined as the change of its position. Depending on the number of objectives considered by the fitness function, the PSO algorithms can be classified as single and multi-objective algorithms [25].

The information available for each particle is based on its own experience and the knowledge of the performance of other particles in its neighborhood. Therefore, each particle adjusts its trajectory based on its own previous best local position and the previous best global position attained by any particle of the swarm, namely p_{id} and p_{gd}. The velocities and positions of particles are updated using Equations (8) and (9), respectively:

$$v_{id}(t+1) = wv_{id}(t) + c_1 rand_1 (p_{id} - x_{id}(t)) + c_2 rand_2 \left(p_{gd} - x_{id}(t)\right) \tag{8}$$

$$x_{id}(t+1) = x_{id}(t) + v_{id}(t) \tag{9}$$

where t is the iteration counter, w is the inertia weigh, c_1 and c_2 are the acceleration coefficients, and $rand_1$ and $rand_2$ are two random numbers uniformly distributed in the interval [0, 1]. The inertia weight controls the impact of previous velocities on the current velocity and it is used to control the convergence of the PSO [12]. To reduce this weight over the iterations allowing the algorithm to exploit some specific areas, w is updated according to the following equation:

$$w = w_{max} - \frac{w_{max} - w_{min}}{iter_{max}} iter \tag{10}$$

where w_{max} and w_{min} are the maximum and minimum values that the inertia weight can take, $iter$ the current iteration of the algorithm and $iter_{max}$ the maximum number of iterations. The acceleration coefficients c_1 and c_2 control how far a particle moves in a single iteration. The velocity update in Equation (8) has three major components. The first one is the inertia, which models the tendency of the particle to continue in the same direction that it has been travelling. The second component is usually referred as memory and it is the linear attraction towards the best position ever found by the given particle p_{id} scaled by a random weight $c_1 rand_1$. The last component, usually referred as cooperation or social knowledge, is the linear attraction towards the best position found by any particle p_{gd}, scaled by another random weight $c_2 rand_2$.

Algorithm 1

Objective function $f(x)$, $x_i = (x_{i1}, x_{i2}, \ldots, x)$
Initialize locations x_i and velocity v_i, $i = 1, 2, \ldots, P$
Find p_{gd} from $\min\{f(x_1), \ldots, f(x_P)\}$ at (t = 0)
While (criterion)
    For loop over all P particles and all d dimensions
        Generate new velocity $v_{id}(t+1)$ using (8)
        Calculate new locations $x_{id}(t+1)$ using (9)
        Evaluate objective function at new locations $x_{id}(t+1)$
        Find p_{id} for each particle x_{id}
    End for
        Find the current p_{gd}
        Update $t = t + 1$
End while
Output the final results x_{id} and p_{gd}
---

with p_{id} the PSO algorithm tries to force exploitation around local optimums, while with p_{gd} the algorithm explores new areas of the search space. Both features are the main tools for the PSO algorithm to achieve satisfactory results in complex optimization problems like the one presented in this work. Algorithm 1 represents the original implementation of the PSO algorithm used in this work. Furthermore, in this work, each individual will represent the set of electrical parameters to be estimated using the PSO algorithm, whose result will be proven to converge to an optimal solution.

3. Suggested Estimation Procedure

The proposal presented in this work utilizes both the standstill technique and the PSO procedure that have been particularized to the system under study, which is a symmetrical five-phase induction machine with distributed windings fed by a two-level VSI. In order to have a better understanding of the estimation procedure, this section will detail the standstill scheme, where an insight into how the electrical parameters are estimated is provided. Then, the application of the search engine based on the PSO method to obtain the final estimation is described.

3.1. Standstill Procedure in Five-Phase Induction Drives

The basis of standstill identification schemes is that the machine model can be simplified when the rotor speed is zero ($\omega_r = 0$). This can be obtained with an appropriate stator winding arrangement that avoids the generation of electrical torque. Several stator winding arrangements can be chosen, generating different stator current components. Table 1 summarizes two winding arrangements proposed in [9] for the identification of the electrical parameters in the α-β (first row) and x-y (second row) subspaces. The first one maximizes the α-axis component with respect to the x-axis component (winding connection 1), while the remaining components are zero. This arrangement allows two identification processes in the α-β subspace for the estimation of the rotor parameters (R_r, L_{lr}) and the magnetizing inductance (L_m). The second one maximizes the x-axis component with respect to the α-axis component (winding connection 2), generating null components in the rest. Then, this second arrangement allows one identification process in the x-y subspace to estimate the stator resistance (R_s) and stator leakage inductance (L_{ls}) parameters. The resulting discrete dynamics models, which will be used in the identification process, are obtained as follows.

Table 1. Two available windings' arrangements in a five-phase induction machine for the single-phase standstill estimation procedure.

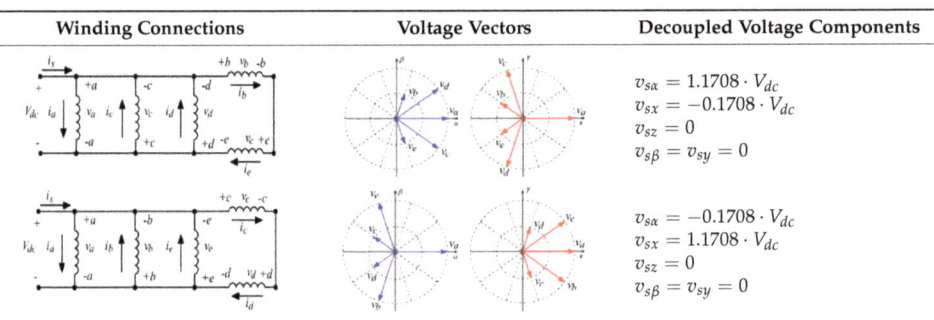

Winding Connections	Voltage Vectors	Decoupled Voltage Components
		$v_{s\alpha} = 1.1708 \cdot V_{dc}$ $v_{sx} = -0.1708 \cdot V_{dc}$ $v_{sz} = 0$ $v_{s\beta} = v_{sy} = 0$
		$v_{s\alpha} = -0.1708 \cdot V_{dc}$ $v_{sx} = 1.1708 \cdot V_{dc}$ $v_{sz} = 0$ $v_{s\beta} = v_{sy} = 0$

The first identification model focuses on the α-β plane and it is shown in the upper row of Table 1. Then, a winding arrangement is chosen to minimize stator voltage in the x-y subspace and reduce any interference between orthogonal frames. However, notice that the obtained stator voltage is not null in the x-y plane, so certain disturbance in the identification process is generated. The stator and rotor current responses in the α–axis can be described by the following equations:

$$V_{s\alpha}(s) = (R_s + sL_s)I_{s\alpha}(s) + sL_m I_{r\alpha}(s) \quad 0 = (R_r + sL_r)I_{r\alpha}(s) + sL_m I_{s\alpha}(s) \tag{11}$$

The transfer function that models the current response in the α–β subspace is as follows:

$$V_{s\alpha}(s) = (R_s + s\sigma L_s)I_{s\alpha}(s) + \frac{sK_T}{1 + s\tau_r}I_{s\alpha}(s) \tag{12}$$

where $K_T = L_m^2/L_r$, $\tau_r = L_r/R_r$ and $\sigma L_s = L_s - K_T$.

The continuous-time transfer function that describes the α–axis stator current response can be simplified using the term $V_{sr}(s)$ detailed in (13), as it is shown in (14), and discretized using a zero-order holder as it is stated in (15):

$$V_{sr}(s) = V_{s\alpha}(s) - (R_s + s\sigma L_s)I_{s\alpha}(s) \tag{13}$$

$$\frac{I_{s\alpha}(s)}{V_{sr}(s)} = \frac{(1 + s\tau_r)}{K_T s} \tag{14}$$

$$\frac{I_{s\alpha}(z)}{V_{sr}(z)} = Z\left\{\frac{1-e^{-sT_s}}{s} \cdot \frac{1+s\tau_r}{K_T s}\right\} = \frac{\tau_r + (T_s + \tau_r)z^{-1}}{K_T(1-z^{-1})} \tag{15}$$

where T_s is the sampling period.

This model, also called "full-order transfer function model in the α-β subspace", provides information of current response in the α-β plane. In essence, the same model has been so far used in the identification process of three-phase machines using standstill techniques whose parameters are identified using this transfer function.

The model in the x-y subspace is now studied. The continuous-time transfer function that describes the x-axis current response is obtained after creating a stator voltage using the winding arrangement shown in the second row in Table 1:

$$\frac{I_{sx}(s)}{V_{sx}(s)} = \frac{1}{(R_s + sL_{ls})} = \frac{1}{R_s(1 + s\tau_{ls})} \tag{16}$$

where $\tau_{ls} = L_{ls}/R_s$.

This model in the x-y subspace can be referred as the "stator leakage inductance model" because it contributes to the estimation allowing the identification of the L_{ls} parameter. The input voltage in the x–axis depends on V_{dc}, as it is detailed in Table 1 (fourth column). Notice that the obtained stator voltage is not null in α-β plane. Therefore, certain disturbance to the identification process is generated as in the previous case. The model of the current response is then discretized using a zero-order hold as follows:

$$\frac{I_{sx}(z)}{V_{sx}(z)} = Z\left\{\frac{1-e^{-sT_s}}{s} \cdot \frac{1}{R_s(1+s\tau_{ls})}\right\} = \frac{\left(1-e^{-T_s/\tau_{ls}}\right)z^{-1}}{R_s(1-e^{-T_s/\tau_{ls}}z^{-1})} \quad (17)$$

The stator leakage inductance model provides additional information, compared with the three-phase case, about the identification of R_s and L_{ls} parameters, and will be used for this purpose.

3.2. Search Engine for the Estimation Process Using PSO

The main idea of using the PSO algorithm in this complex application is to converge towards a good solution of the estimated electrical parameters of a five-phase induction machine. Each particle is composed of a set of electrical parameters like an unknown vector $x = [R_s\ R_r\ L_m\ L_{ls}\ L_{lr}]$ to be accurately estimated. The fitness function used to evaluate the quality of every particle in the population is the mean squared error (MSE) between the outputs given by the real system (the multiphase induction machine, y_α and y_x in Equation (18) and the outputs given by a modelled system (using Matlab and named $\hat{y}_\alpha$ and $\hat{y}_x$). Both systems (the real machine and the Matlab-based model) are governed using stator voltages in the standstill configuration to generate first a response in the α-β subspace and then in the x-y plane. The full-order model is avoided to guarantee that the estimation of the α-β parameters (involved in the main control magnitudes of the electrical drive such as the electrical torque and the stator flux production) is made without having any interference of the x-y plane, which is related to the electrical losses in a machine with distributed winding. For this reason, the same weights have been considered for both subspaces α-β and x-y. Consequently, the proposed fitness function g for this study is defined as follows:

$$g = \sqrt{MSE_x^2 + MSE_\alpha^2}$$

$$MSE_\alpha^2 = \frac{1}{N_\alpha}\sum_{k=1}^{N_\alpha}\|y_\alpha(k) - \hat{y}_\alpha(k)\|^2 \quad (18)$$

$$MSE_x^2 = \frac{1}{N_x}\sum_{k=1}^{N_x}\|y_x(k) - \hat{y}_x(k)\|^2$$

where the MSE_α and MSE_x values are the mean squared errors computed for the response in the α- and x-axis, respectively, and N_α and N_x regulate the desired accuracy in the estimation of the α-β and x-y parameters (in this case, the same accuracy has been selected).

The complexity of the estimation procedure comes from adjusting simultaneously the two regression models of α-β and x-y planes based on the response of the multiphase machine in standstill arrangements to known input signals. On the one hand, the regression model of α-β plane that allows the estimation of R_r, L_{lr} and L_m parameters. On the other hand, the x-y plane that enables the estimation of R_s and L_{ls}. Notice that the five electrical parameters have continuous values ranging from the intervals included in Table 3 (see Section 4.1 for more details). Therefore, the complexity of the optimization problem consists in finding the most optimal values that reduce the error among the simulated response of electrical machine model with the electrical parameters as inputs and the real response obtained from experiments.

4. Experimental Assessment

The performance of the proposal is analyzed using an experimental test bench based on a symmetrical five-phase induction machine with distributed windings. The multiphase machine was built from a commercial three-phase induction machine that has been rewound and reassembled. Then, the proposed estimation technique is applied to obtain the unidentified vector $x = [R_s\ R_r\ L_m\ L_{ls}\ L_{lr}]$ that represents the electrical parameters of the multi-phase machine. Figure 3 shows an scheme of the experimental test bench, where pictures of electronic equipment are included. The VSI-based multiphase power converter is built from two commercial three-phase modules from Semikron (SKS21F) that are linked to a unique DC of up to 300 V. The controller is based on a well-known digital signal processor from Texas Instruments (12500 TI Boulevard, Dallas, TX, USA) and Technosoft (Avenue des Alpes 20, 2000 Neuchâtel, Switzerland), the TMS320LF28335 and the MSK28335 board, respectively. Sensing some electrical variables (stator currents and voltages) is a major requirement in the estimation strategy, which it is done using two different sensors from LEM (Chemin des Aulx 8, P.O. Box 35, 1228 Plan-les-Ouates, Switzerland), the LA-55P and LV-25P devices. It is important to highlight that the voltage electrical signals obtained from the sensors are filtered using analog low-pass filters with a cut-off frequency of 1.5 kHz. It is also interesting to remark that the windings of the multiphase machine must be rearranged to avoid torque generation and to assure the standstill behavior. This is done following the connection scheme shown in the first row of Table 1.

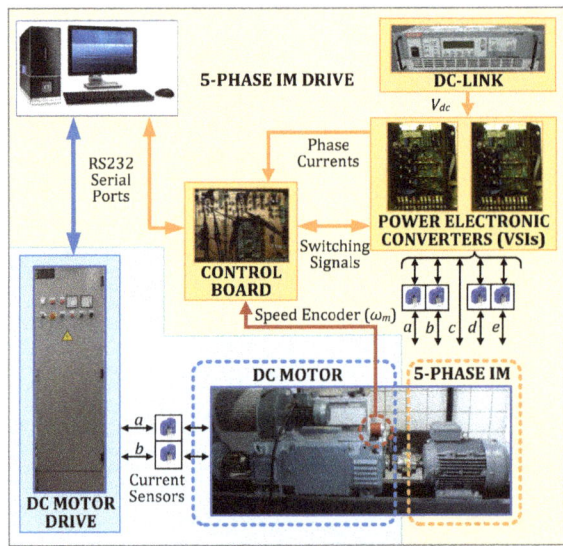

Figure 3. Scheme of the experimental test bench.

4.1. Identification of the Electrical Parameters of the System

Two different stator voltages are applied using the proposed standstill tests in α-β and x-y subspaces and the current responses of the system are recorded. The winding connection shown in the first row of Table 1 is initially used and a step voltage from −20 to 20 V is applied to the machine, Figure 4a, to obtain the current response in the α-axis shown in Figure 4b. This voltage excites the electromagnetic circuit and rotor time constants at standstill in the α-axis, as it is detailed in Equation (15). The winding connection shown in the second row of Table 1 is then used and a three-level signal (−60, 0 and 60 V) with a fundamental frequency of 25 Hz is applied to the stator, see Figure 4c. This stator voltage excites the stator electromagnetic circuit detailed in Equation (17), producing the stator current in the x-axis shown in Figure 4d. The obtained stator current responses y are then compared with the modelled responses $\hat{y}$, evaluated with Equations (15) and (17) in order to compute the fitness function g (18) of each individual in the PSO algorithm.

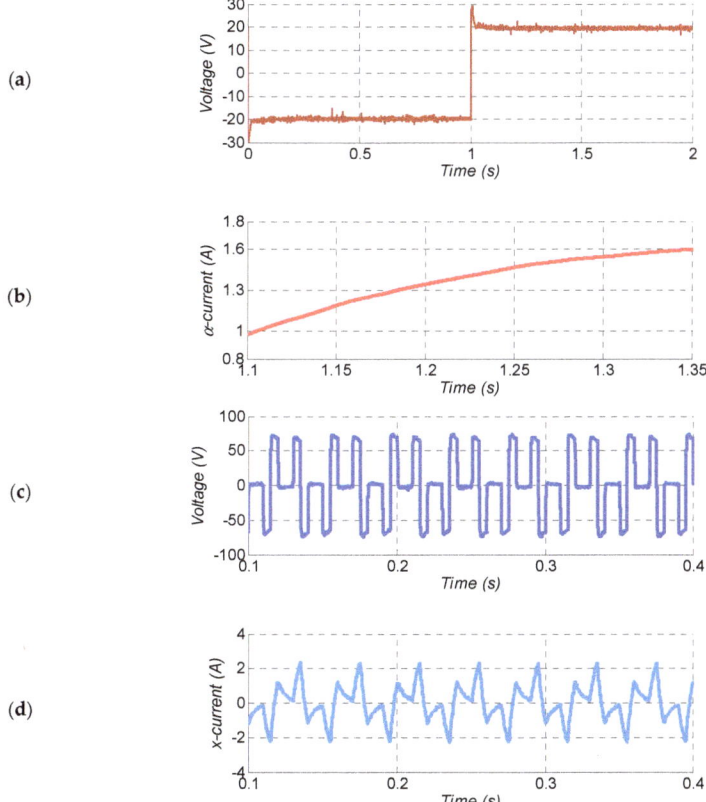

Figure 4. Applied stator voltage in the winding connection 1 (**a**) and the measured $i_{s\alpha}$ (**b**). Applied stator voltage in the winding connection 2 (**c**) and the measured i_{sx} (**d**).

Table 2 contains the configuration parameters used in the PSO algorithm, which is designed to stop under two circumstances. The PSO algorithm has been run for 30 independent trials. Each trial is stopped under the following events: first, if 400 iterations are reached, or second, if the best global position does not change during 40 iterations or the change is lower than the lowest error gradient tolerance (*errgrad*). Notice that realistic variation intervals for the electrical parameters of the machine (summarized in Table 3) must be supplied to the PSO algorithm to ensure a proper solution. Consequently, previous knowledge of the real system is required to apply the proposed estimation algorithm.

Table 2. Configuration parameters of the PSO algorithm.

PSO Parameter	Value
Number of trials	30
Number of particles (ps)	[25, 125]
Acceleration coefficients (c_1 and c_2)	$c_1 = c_2 = [0.1, 2]$
Inertia weights (w_{max} and w_{min})	$w_{max} = [0.5, 1.4]$ and $w_{min} = 0.3$
Maximum particle velocity (v_{max})	$v_{max} = [1, 3]$
Lowest error gradient tolerance (*errgrad*)	$errgrad = 1 \cdot \times 10^{-6}$
Maximum number of generation without error change (*errgraditer*)	$errgraditer = 40$
Maximum number of iterations ($iter_{max}$)	$iter_{max} = 400$

Table 3. Parameter ranges for the PSO variables.

Machine Parameter	Interval
R_s (Ω)	[10, 25]
R_r (Ω)	[1, 10]
L_m (H)	[0.5, 0.7]
L_{ls} (H)	[0.010, 0.160]
L_{lr} (H)	[0.010, 0.060]

To select the suitable values of the adjusting parameters of the PSO algorithm, massive simulations have been conducted varying the parameters, such as ps, c_1, c_2, V_{max}, and W_{max}, according to the intervals included in Table 2. A grid search has been conducted by dividing each interval of each configuration parameter into four. Each point of the grid has been evaluated for 30 independent trials. Therefore, about 8000 simulations have been conducted. In general, the results are satisfactory for all the cases considered since important differences in the obtained results are not observed. According to the results in Table 4, the most suitable adjusting parameters for the PSO implementation are: $ps = 75$, $c_1 = c_2 = 1$, $V_{max} = 1$, and $W_{max} = 0.9$. Table 4 details the estimated parameters for the best run, obtained with a computed error of 0.1701. The identification method based on gradient-based optimization algorithms and proposed in [9] was also applied to compare with these results, giving an estimated error about 2.58% higher than the obtained using the PSO technique. Then, an improvement in the estimation procedure is obtained, which proves the interest and applicability of the proposal. Notice that the accuracy of the electrical parameters has strong impact on the closed-loop performance of the system, being an important trend in control theory for electrical drives.

Table 4. Obtained parameters using the PSO algorithm.

Machine Parameter	Value
R_s (Ω)	19.4462
R_r (Ω)	6.7659
L_m (H)	0.6565
L_{ls} (H)	0.1007
L_{lr} (H)	0.0386

Moreover, Table 5 includes statistical results with respect to the number of particles used in the PSO algorithm. Notice that important differences are not noticeable when the number of particles is higher than 75. Therefore, it has been chosen 75 as appropriated number of particles for the target optimization problem.

Table 5. Estimation error versus number of particles ps.

Ps	Max.	Mean	Std.
25	0.3340	0.3450	8.4569×10^{-6}
50	0.1879	0.1928	3.4569×10^{-6}
75	0.1701	0.1745	2.4569×10^{-6}
100	0.1722	0.2038	2.2269×10^{-6}
125	0.1725	0.1755	2.1100×10^{-6}

4.2. Stadistical Analysis and Comparison with a Grandient-Based Approach

Figure 5 depicts several performance metrics of the proposed approach. Figure 5a shows the boxplot for the error distribution obtained in the conducted trials, where the obtained distribution data results are shown in a standardized way. Boxplots are normally used to show the dispersion of the simulation results. They represent the median and the 25% and 75% of the simulation results. Therefore, boxplots are ideal tool for statistical analysis. It may be observed that the deviation in the distribution of the results is very low (2.4569 × 10^{-6}) in comparison with the evaluated mean. Figure 5b compares the obtained average results using the proposed PSO-based technique and the gradient-based approach presented in [8,9]. It may be concluded that the proposed estimation technique reduces the error value considerably. Another important issue to highlight is that the 95% confident interval for the obtained results using the PSO-based approach is 8.1786 × 10^{-6}. Therefore, the proposed PSO-based approach clearly outperforms the gradient-based technique for all the conducted trials. Finally, Figure 5c,d depict the performance of the proposed estimation technique in terms of execution time and number of generations required for convergence. Notice that the execution time is not critical in this optimization problem since it is obviously and offline procedure. Nevertheless, the proposed approach provides results in 4000 s (approximately 1 h) on average. These results were obtained using a Toshiba Satellite L755 Intel®Core™ i7 2670QM, 4 G RAM. Consequently, the execution time can be considerably reduced using a modern workstation. Regarding the number of generations required for the convergence, it may be observed (see Figure 5d) that the convergence is usually reached in less than 100 generations. The considered stopping criterion in the PSO configuration (until 400 generations can be run if necessary) is then a suitable configuration set.

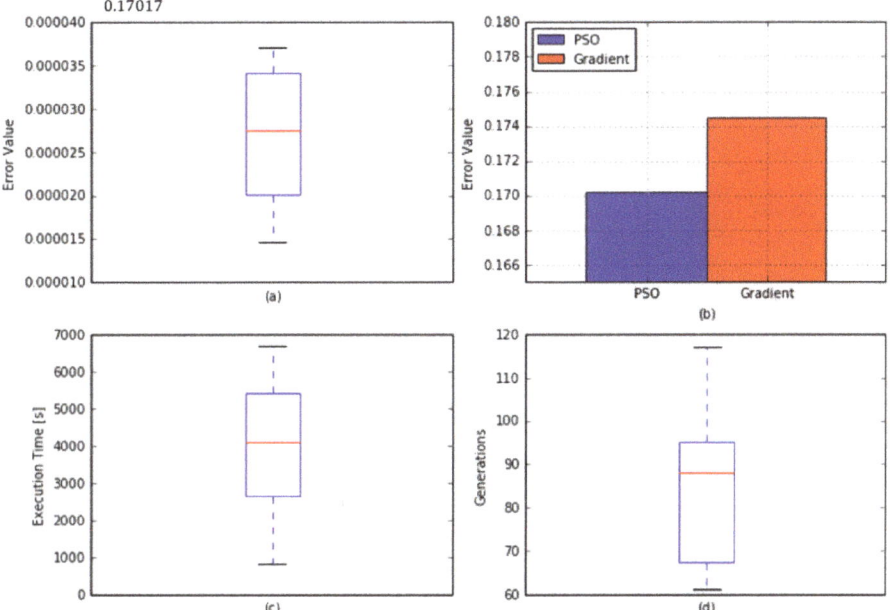

Figure 5. Statistical analysis of the results showing (**a**) the obtained estimation error, (**b**) a performance comparison between the PSO-based and gradient-based approaches, (**c**) the execution time of the proposed estimation procedure, and (**d**) the number of generation required for the convergence of the algorithm.

4.3. Experimental Validation of the Estimated Parameters

To analyze the validity of the results obtained, a graph of the estimated transfer functions is plotted with a log-frequency axis in order to compare the theoretical and experimental frequency responses of the system (Bode plot). The mathematical representation of the system shown in previous equations is compared with the real behavior using the proposed winding arrangements and the estimated parameters. Figures 6 and 7 show the obtained results. Figure 6 depicts theoretical and experimental Bode plots in the α-β plane in blue and red ink, respectively. A good agreement is observed in Figure 6. Moreover, Figure 7 shows theoretical and experimental Bode plots in the x-y plane. Again, theoretical and experimental behaviors are quite similar. Notice that some differences exist. These differences can be justified due to the inaccuracy of the initial modelling assumptions, the error in the measurement process and the relationship with the frequency of the electrical parameters of the machine. Such differences produce that the experimental transfer function varies from the theoretical one as the frequency increases.

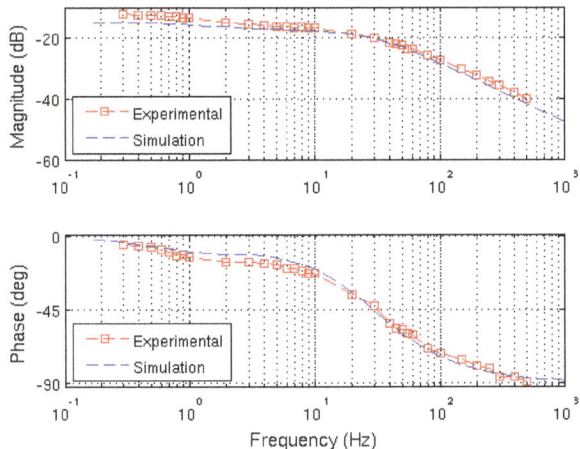

Figure 6. Simulation (blue dashed plot) and experimental (red squares plot) Bode frequency responses in the α-β subspace.

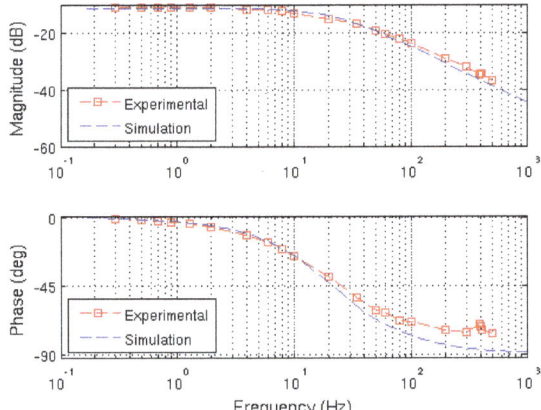

Figure 7. Simulation (blue dashed plot) and experimental (red squares plot) Bode frequency responses in the x–y subspace.

5. Conclusions

This paper describes a novel off–line procedure for the estimation of the electrical parameters of a multi-variable electro-mechanical system. Unlike recently proposed gradient-based methods, this proposal utilizes the PSO technique as a proof-of-concept of the application of meta-heuristic optimization algorithms in the estimation of electrical parameters based on standstill methods. The method has been tested in a real system using a multiphase test rig with a five-phase induction machine. In addition, it has been compared with sinusoidal and time domain gradient-based estimation techniques. A reduction higher than 2.58% is obtained in the best solution error. Notice that this result is relevant for the development of high performance modern controllers where the knowledge of the electrical parameters of the multiphase drive is crucial, such as in predictive control algorithms. Furthermore, this work paves the way for future application of other variants of population-based techniques, such as the genetic algorithm (GA) and firefly algorithm (FA), and other recent trajectory-based algorithm, such as simulated annealing (SA), tabu search (TB) and harmony search (HS), among others, for the optimization problem presented in the estimation of electrical parameters of multiphase machines based on standstill tests. It is interesting to note that although the proposal has been tested for a particular multiphase drive (five-phase machine), it can be extended for identification purposes in different electrical drives. There are no restrictions in the application of our proposal in different multiphase electrical machines, although the propose windings' arrangement (Table 1) is no longer valid and must be adapted to the standstill requirements of the new machine.

Author Contributions: Conceptualization, F.B. and S.L.T.; Methodology, F.B., J.R. and D.G.-R.; Software, D.G.-R. and J.R.; Validation, D.G.-R., J.R. and I.G.-P.; Formal Analysis, D.G.-R., J.R. and I.G.-P.; Investigation, D.G.-R. and J.R.; Resources, F.B. and M.J.D.; Data Curation, D.G.-R. and J.R.; Writing-Original Draft Preparation, F.B. and D.G.-R.; Writing-Review & Editing, F.B., D.G.-R., I.G.-P., M.J.D.; Visualization, F.B.; Supervision, F.B.; Project Administration, F.B.; Funding Acquisition, F.B.

Funding: This research received no external funding.

Acknowledgments: The authors would like to thanks the University of Seville and the Spanish national R + D + I program (under references DPI2013-44278-R and DPI2016-76144-R) for the equipment used for experiments.

Conflicts of Interest: The authors declare no conflict of interest.

References

1. Barrero, F.; Duran, M.J. Recent Advances in the Design, Modeling and Control of Multiphase Machines—Part 1. *IEEE Trans. Ind. Electron.* **2016**, *63*, 449–458. [CrossRef]
2. Duran, M.J.; Barrero, F. Recent Advances in the Design, Modeling and Control of Multiphase Machines—Part 2. *IEEE Trans. Ind. Electron.* **2016**, *63*, 459–468. [CrossRef]
3. Duran, M.J.; Riveros, J.; Barrero, F.; Guzmán, H.; Prieto, J. Reduction of common-mode voltage in five-phase induction motor drives using predictive control techniques. *IEEE Trans. Ind. Appl.* **2012**, *48*, 2059–2067. [CrossRef]
4. Riveros, J.A.; Barrero, F.; Levi, E.; Durán, M.J.; Toral, S.; Jones, M. Variable-speed five-phase induction motor drive based on predictive torque control. *IEEE Trans. Ind. Electron.* **2013**, *60*, 2957–2968. [CrossRef]
5. Carraro, M.; Zigliotto, M. Automatic Parameter Identification of Inverter-Fed Induction Motors at Standstill. *IEEE Trans. Ind. Electron.* **2014**, *61*, 4605–4613. [CrossRef]
6. Lee, S.H.; Yoo, A.; Lee, H.-J.; Yoon, Y.-D.; Han, B.-M. Identification of Induction Motor Parameters at Standstill Based on Integral Calculation. *IEEE Trans. Ind. Appl.* **2017**, *53*, 2130–2139. [CrossRef]
7. He, Y.; Wang, Y.; Feng, Y.; Wang, Z. Parameter Identification of an Induction Machine at Standstill Using the Vector Constructing Method. *IEEE Trans. Power Electron.* **2012**, *27*, 905–915. [CrossRef]
8. Yepes, A.G.; Riveros, J.A.; Doval-Gandoy, J.; Barrero, F.; López, O.; Bogado, B.; Jones, M.; Levi, E. Parameter identification of multiphase induction machine with distributed windings—Part 1: Sinusoidal excitation methods. *IEEE Trans. Energy Convers.* **2012**, *27*, 1056–1066. [CrossRef]
9. Riveros, J.A.; Yepes, A.G.; Barrero, F.; Doval-Gandoy, J.; Bogado, B.; Lopez, O.; Jones, M.; Levi, E. Parameter identification of multiphase induction machine with distributed windings—Part 2: Time domain techniques. *IEEE Trans. Energy Convers.* **2012**, *27*, 1067–1077. [CrossRef]
10. Sakthivel, V.P.; Bhuvaneswari, R.; Subramanian, S. Artificial immune system for parameter estimation of induction motor. *Expert Syst. Appl.* **2010**, *37*, 6109–6115. [CrossRef]
11. Bettayeb, M.; Qidwai, U. A hybrid least squares-GA-based algorithm for harmonic estimation. *IEEE Trans. Power Deliv.* **2003**, *12*, 377–382. [CrossRef]
12. Wang, D.; Tan, D.; Liu, L. Particle swarm optimization algorithm: An overview. *Soft Comput.* **2018**, *22*, 387–408. [CrossRef]

13. Fernandez-Martinez, J.L.; Garcia-Gonzalo, E. Stochastic Stability Analysis of the Linear Continuous and Discrete PSO Models. *IEEE Trans. Evol. Comput.* **2011**, *15*, 405–423. [CrossRef]
14. Del Valle, Y.; Venayagamoorthy, G.K.; Mohagheghi, S.; Harley, R.G.; Hernandez, J.C. Particle swarm optimization: Basic concepts, variants and applications in power systems. *IEEE Trans. Evol. Comput.* **2008**, *12*, 171–195. [CrossRef]
15. Eslami, M.; Shareef, H.; Mohamed, A.; Khajehzadeh, M. An efficient particle swarm optimization technique with chaotic sequence for optimal tuning and placement of PSS in power systems. *Int. J. Electr. Power Energy Syst.* **2012**, *43*, 1467–1478. [CrossRef]
16. Guzman, H.; Duran, M.J.; Barrero, F.; Zarri, L.; Bogado, B.; Prieto, I.G.; Arahal, M.R. Comparative Study of Predictive and Resonant Controllers in Fault-Tolerant Five-phase Induction Motor Drives. *IEEE Trans. Ind. Electron.* **2016**, *63*, 606–617. [CrossRef]
17. Zhang, J.; Yang, S. A novel PSO algorithm based on an incremental-PID-controlled search strategy. *Soft Comput.* **2016**, *20*, 991–1005. [CrossRef]
18. Rada-Vilela, J.; Johnston, M.; Zhang, M. Population statistics for particle swarm optimization: Single-evaluation methods in noisy optimization problems. *Soft Comput.* **2015**, *19*, 2691–2716. [CrossRef]
19. Yıldız, A.R. A novel particle swarm optimization approach for product design and manufacturing. *Int. J. Adv. Manuf. Technol.* **2009**, *40*, 617–628. [CrossRef]
20. Yıldız, A.R. A new hybrid particle swarm optimization approach for structural design optimization in automotive industry. *J. Automob. Eng.* **2012**, *226*, 1340–1351. [CrossRef]
21. Yıldız, A.R. Comparison of evolutionary based optimization algorithms for structural design optimization. *Eng. Appl. Artif. Intell.* **2013**, *26*, 327–333. [CrossRef]
22. Sheikhan, M.; Mohammadi, N. Time series prediction using PSO-optimized neural network and hybrid feature selection algorithm for IEEE load data. *Neural Comput. Appl.* **2013**, *23*, 1195. [CrossRef]
23. Krama, A.; Zellouma, L.; Rabhi, B.; Refaat, S.; Bouzidi, M. Real-Time Implementation of High Performance Control Scheme for Grid-Tied PV System for Power Quality Enhancement Based on MPPC-SVM Optimized by PSO Algorithm. *Energies* **2018**, *11*, 3516. [CrossRef]
24. Yun, P.; Ren, Y.; Xue, Y. Energy-Storage Optimization Strategy for Reducing Wind Power Fluctuation via Markov Prediction and PSO Method. *Energies* **2018**, *11*, 3393. [CrossRef]
25. Yildiz, A.R.; Solanki, K.N. Multi-objective optimization of vehicle crashworthiness using a new particle swarm based approach. *Int. J. Adv. Manuf. Technol.* **2012**, *59*, 367–376. [CrossRef]

© 2019 by the authors. Licensee MDPI, Basel, Switzerland. This article is an open access article distributed under the terms and conditions of the Creative Commons Attribution (CC BY) license (http://creativecommons.org/licenses/by/4.0/).

MDPI
St. Alban-Anlage 66
4052 Basel
Switzerland
Tel. +41 61 683 77 34
Fax +41 61 302 89 18
www.mdpi.com

Energies Editorial Office
E-mail: energies@mdpi.com
www.mdpi.com/journal/energies

www.ingramcontent.com/pod-product-compliance
Lightning Source LLC
LaVergne TN
LVHW071956080526
838202LV00064B/6760